WAKEFIELD PRESS

BUSH MECHANICS

BUSH MECH

FROM YUENDUMU TO THE WORLD

ANTICS

Edited by Mandy Paul and Michelangelo Bolognese

Wakefield Press

Wakefield Press
16 Rose Street
Mile End
South Australia 5031
www.wakefieldpress.com.au

First published 2017
Reprinted 2018, 2022

Edited by Margot Lloyd, Wakefield Press
Designed by Liz Nicholson, designBITE
Typeset by Clinton Ellicott, Wakefield Press
Printing and quality control in China by Tingleman Pty Ltd

ISBN 978 1 74305 515 1

 A catalogue record for this
book is available from the
National Library of Australia

This project has been assisted by the Australian
Government's Visions of Australia program.

CONTENTS

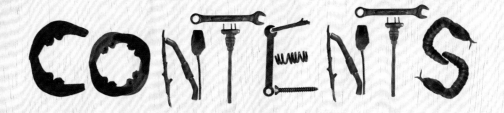

Acknowledgements

This catalogue and the exhibition it accompanies are the result of a partnership between the National Motor Museum (a museum of the History Trust of South Australia) and Pintubi Anmatjere Warlpiri (PAW) Media.

Our heartfelt thanks go to the board and staff of PAW Media for supporting the project, and particularly to board chair Francis Jupurrurla Kelly. Thanks also to David Batty for his enthusiastic response to the many requests that have come his way over the last year. Special thanks are due to Jeff Bruer for liaison and assistance with locating images in the PAW Media archives. Thanks too to Jason Japaljarri Woods for the artwork featured in the exhibition and this catalogue, including his new version of the *Bush Mechanics* alphabet developed especially for the exhibition.

Images used in this book were also sourced from the Australian Institute of Aboriginal and Torres Strait Islander Studies, the National Library of Australia, the National Film and Sound Archive, the National Museum of Australia and the State Library of New South Wales. Photography of *Bush Mechanics: The exhibition* installed at the National Motor Museum is by Andre Castellucci. Thanks to Suzanne Redman for assistance with image production, and to Corinne Ball and Nikki Sullivan for transcription.

This catalogue was inspired by the work of Melinda Hinkson and Georgine Clarsen, and we are delighted to have been able to include chapters by both.

Thanks to Michael Bollen and the team at Wakefield Press for transforming the manuscript into a thing of beauty.

Finally, we would like to acknowledge the financial assistance received from the Australian Government's Visions of Australia program. Neither the exhibition tour nor this publication would have been possible without this assistance.

Aboriginal and Torres Strait Islander readers are warned that this
book contains images of people who are deceased.

Nyampuju yapa-patu-kulu. Panukari kalu-juku marda nyinami.
Panukari mardalu lawa-nyinajalku.

Foreword

A short low-budget television production at the turn of the new millennium threw a spotlight on the Warlpiri bush mechanics from Australia's Tanami Desert and drew millions of admirers. The inventiveness and technological bravado of the tricks they used to overcome mechanical adversity in Central Australia became as celebrated as their travels and adventures. The *Bush Mechanics* exhibition is a long overdue addition to the National Motor Museum's business of telling Australia's motoring history as we explore automobility in Aboriginal Australia and continue our journey of recognising and telling the untold stories of Australian motoring.

This catalogue uncovers more of the relationship between Aboriginal Australians and the motor car than can be found in the exhibition, which is rich in interactivity and visual content but deliberately light on text. Through Georgine Clarsen's 'Bush Mechanics: Then and now', we delve into the big themes that *Bush Mechanics* offers and the radically different view into Aboriginal Australia provided by the series – and now the exhibition. Georgine also provides the reader with a deeper understanding of Warlpiri car culture and history, while weaving in

The EJ Holden in the *Bush Mechanics* series. National Film and Sound Archive of Australia 135784.

wider Aboriginal history and connection to vehicles, which can be traced back to the very early days of motoring in Australia.

Melinda Hinkson's "'We have always moved around'" provides a further journey into Warlpiri culture and history. Melinda explores the impact of colonisation on Tanami Desert dwellers and tells the story of the establishment of Yuendumu. The history of Warlpiri's embrace of media brings further insight into the evolving nature of Aboriginal storytelling and communication. The establishment of Warlpiri Media Association (now Pintubi Anmatjere Warlpiri or PAW Media), the makers and producers behind *Bush Mechanics*, and the power of 'media tools' in Warlpiri hands is also told in Melinda's chapter. She explores the distinctive style of Warlpiri media makers as well as the 'community making' power of the media repurposed in Aboriginal hands.

The director and writer of *Bush Mechanics*, David Batty, and co-director and star, Francis Jupurrurla Kelly have been working together for three decades producing media in the bush. In the chapter that follows, they tell how the concept for *Bush Mechanics* was born from a shared passion for improvised mechanical repairs. And the repairs themselves – *nyurulypa*, in Warlpiri – are the subject of the next chapter by the exhibition's curator, Michelangelo Bolognese.

The catalogue concludes with the story of how and why the National Motor Museum and the History Trust of South Australia embarked on an exhibition about the *Bush Mechanics* television series. Michelangelo has often recounted his introduction to Australian life, when, as a young teen from Italy, he turned his television on for the first time and landed on the quirky series about Aboriginal men and their cars, unlike anything he had seen before. In '*Bush Mechanics*: The exhibition', he and project manager Mandy Paul provide an entertaining insight into the making of the exhibition, while also demonstrating the role a museum can play in taking an audience on a new journey through what was thought to be familiar country.

The National Motor Museum is a museum of the History Trust of South Australia. The development of *Bush Mechanics: The exhibition* in partnership with PAW Media has been an important step for the museum. We have enjoyed watching it develop from a concept into the informative yet slightly irreverent exhibition we have today, very much in the spirit of the original series. Visitors to the museum have delighted in reliving the original series through a new lens, this time not of the filmmaker, but of the historian and curator. This catalogue provides yet another lens. We hope you enjoy the view.

Paul Rees

Director, National Motor Museum

Making *Bush Mechanics*. Pintubi Anmatjere Warlpiri Media.

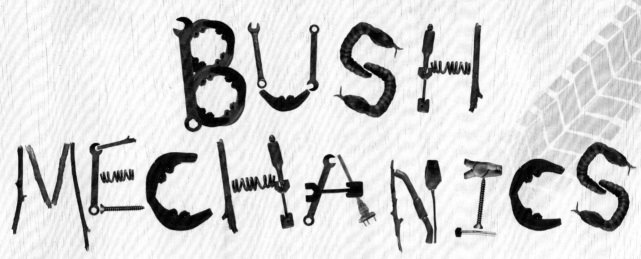

BUSH MECHANICS

Then and now

Georgine Clarsen

Georgine Clarsen reminds us that there have been Aboriginal bush mechanics in Australia for at least 100 years. She shows how the mechanical creativity so energetically displayed by men from Yuendumu in the Bush Mechanics *series is the culmination of decades of adapting automobiles to suit the cultural and material realities of Aboriginal people's lives under colonisation.*

Brake fluid made from laundry powder and water? Welding a broken muffler with jumper leads, fencing wire and a car battery? New brake pads carved from mulga wood with a tomahawk, or an emergency clutch plate shaped out of an old boomerang? How about windscreen wiper blades made from strips of blanket? Such is the car repair advice offered by the heroes of the *Bush Mechanics* series. Presented with humour, as well as a large dose of self-parody, theirs is mechanical advice like no other. *Nyurulypa* ('good tricks') are what these bush mechanics from Yuendumu in the Central Desert call it. *Nyurulypa* is specialised knowledge that has been hard won in the collective experience of a particular kind of *mutikar* culture.

Maverick acts of resuscitation performed on bush bombs by the mechanics from Yuendumu are far from the usual automobile stories. The cars we see in this series seem only barely related to the alphabet magic of contemporary automobile production. No ABS, EFI, or all-mode 4WD in the world of *Bush Mechanics*. And the robustly practical *nyurulypa* the men offer have no parallel in the reverential approach to cars found in the television lifestyle programs more often associated with automobile advice. *Mutikars* for our teachers of *nyurulypa* are certainly about pleasure, excitement, freedom and untrammelled mobility, as promoted by car advertisements. But in *Bush Mechanics* the pleasures of these *yapa* (as Warlpiri refer to themselves) men's engagements with cars come in a very different form – greasier and harsher, collective rather

Unnamed man in his Ford, 1920s. Photographer: Herbert Basedow. National Museum of Australia.

than individualistic, and with only a tangential relationship to 'the' economy.

David Batty and Francis Jupurrurla Kelly were the joint directors of the first *Bush Mechanics* episode in 1998. Batty, a non-Indigenous filmmaker who grew up in Wollongong, has been making television programs for and with Aboriginal people for more than thirty years. Kelly is a Warlpiri man who had a leading role in early television production at Yuendumu. It is Kelly's creativity and enthusiasm as a bush mechanic that drives the series. The two worked together on a number of earlier television programs – most notably *Manyu Wana*, a ten-part children's series (the 'Warlpiri Sesame Street'), which began production in 1990.

In the subsequent *Bush Mechanics* episodes that were produced three years after the first, David Batty is credited as director, scriptwriter and cinematographer, with Francis Jupurrurla Kelly credited as actor and co-director. The main onscreen performers, a close-knit gang of five Yuendumu men who were the bush mechanics in the 1998 video, continued their roles in the subsequent series except for the replacement of Adrian Nelson by his brother. All five men – Steven Jupurrurla Morton, Errol Jupurrurla Nelson, Simeon Jupurrurla Ross, Junior Jupurrurla Wilson, and Randall Jupurrurla Wilson – play themselves in dramatised 'real life' stories, in a quasi-documentary format. The production crew was expanded for the series by bringing in Jeni McMahon as producer and Kath Shelper as production manager. Editor of the original episode, David Nixon, remained with the series. New technologies, such as desktop editing software, meant that much of the post-production could take place on location at Yuendumu. That allowed greater community consultation and involvement in the final product. The soundtrack for each episode was recorded at Yuendumu, in the distinctive Central Desert style first brought to outside audiences in the early 1980s by Papunya's Warumpi Band.

Shooting the *Bush Mechanics* series. National Film and Sound Archive of Australia 135691.

Bush Mechanics expresses the fun of roaring around country in clapped-out Fords or (preferably) Holdens, a rifle ready for a shot at some bush food. The men's talent for telling stories is utterly compelling as they reveal their lives with, in, and under their cars. As young men, they do not have access to the car of choice in Central Australia, the robust but expensive Toyota LandCruiser (or 'Troopie') and the series registers *yapa* pride in the rough and ready inventiveness that secures their mobility in spite of their scant finances. It celebrates family and friendship, traditional knowledge applied in a contemporary setting, and their community's long history of mechanical creativity. Contra to Philip Ruddock's infamous declaration in 2000

Improvising, with David Batty on the Bolex. Pintubi Anmatjere Warlpiri Media.

that Aboriginal people were deficient because they had not invented the wheel, these men show that there are many ways to use and value a technology that white Australians presume is their own. These bush mechanics have indeed re-invented the wheel.

Bush Mechanics isn't simply about ingenious ways of keeping cars moving, though the first episode takes that as its major theme. The storyline of that original episode is simple – the five heroes head down the Tanami Road from Yuendumu to buy a car in Alice Springs, 300 km away. The car they are travelling in breaks down, system by system. For each breakdown the men improvise a solution. A puncture is repaired by lifting the car onto a jerry can, removing the wheel nuts with a pair of pliers, and stuffing the tyre casing with spinifex ('let's keep the tube, we might meet some whitefellas with patches'). The dead battery is recharged by heating it beside a fire. The windscreen washer is connected into the fuel line to replace a faulty petrol pump. Finally, with no windscreen, the gearbox operating only in reverse, the driver pumping the washer button on the broken indicator stalk, and with a tyre-less rear wheel spraying an arc of red dirt, the five men inch their way backwards toward the main road to Alice Springs, trying to enjoy the ride. When the car gives up on them altogether, they pile out and give it a desultory kick ('This heap of metal has been nothing but trouble') and head off with their blankets and

Recreating the mulga clutch. Pintubi Anmatjere Warlpiri Media.

a box of matches to find a lift into town. In an Alice Springs car sales yard their bravado melts away and the men choose a car mainly on the basis that it starts and the radio works. They leave town quickly, returning triumphantly to Yuendumu in their new car.

That story – a hilarious sequence of everyday misadventure, told with ironic self-deprecation – is cut with segments in which a number of older men speak directly to the camera, explaining something of their passion for cars. Jack Jakamarra Ross offers a beautifully delivered exposition of why *yapa* prefer old Fords and Holdens rather than new ones. Unlike new models, old cars can be easily understood and they can be repaired with materials lying around anywhere. He shows a fuse holder with brass .303 rifle shells replacing the missing fuses, and points out with a smile that this cannot be done with electronic control systems. His stories bridge old and new practices: 'That's the way we learn. From old peoples that have been driving cars like this – Holden Toranas – for a long time.' The chief bush mechanic and co-director, Francis Jupurrurla Kelly, is also periodically cut into this episode, proudly demonstrating the numerous good tricks that keep *yapa* cars moving. A repair ramp may be built by resting a sapling against an anthill; a tail shaft is easily replaced by rolling a car on its side; timber and fencing wire can fix a broken spring; a door that won't stay closed can be sorted by knocking a wooden peg, latch-like, through the roof channel. Needless to say, these are not the kind of solutions that a health and safety committee would endorse.

As engaging as such creative responses to mechanical misfortune are they are not enough to sustain a longer series. After all, how many ways can the men coax a car back to life, and how many times can an audience be convinced that a piece of mulga wood really will fix the brakes? These sagas of resourcefulness, however, provide a vehicle for the expression of broader contemporary realities for people from the Central Desert, and they are further explored in the subsequent four episodes. That later series turns away from realist modes of narration to pursue other stories in which cars are not always the centre of attention. They tackle larger themes, while continuing to provide plenty of scope for *mutikar* business.

In the first of the later episodes, the bush mechanics form a rock band and are hired to drive to a neighbouring community to play for the kids. After a series of mechanical disasters, they (and their sound equipment) arrive just in time to perform for a happy crowd and collect the promised payment. In the second episode, a nephew who has served his sentence phones the bush mechanics from the Alice Springs gaol and asks them to pick him up. On their way home two of the bush mechanics are themselves arrested for outstanding warrants. Members of a rival football team steal a car from Yuendumu in the third episode, and the bush mechanics chase them back to their community ('they're rubbish that mob'), eventually torching the stolen car. And in the final episode Jangala, the rainmaker, sends them west into saltwater country to collect the pearl shells he needs for drought-breaking ceremonies. Jangala's rituals prove so effective they cause flooding throughout Australia, to the consternation of Monty, Northern Territory television's popular weatherman.

In these four stories, historical, contemporary and mythical characters become blurred, and elements of magical realism are brought into play. Spirit guardians enter plotlines. The chief bush mechanic, dressed in blue overalls, possesses supernatural powers to arrive instantaneously, seemingly in reference to the style of the Japanese television cult classic, *Monkey*, to help the stranded men – though on another occasion he is delivered by tourists in a minibus. The prison episode is built around the simultaneous claim of two laws. The bush mechanics' nephew is brought home from Alice Springs, where he has served a sentence under white law, to face the tribal payback that the old people in his dreams have foreshadowed. In the final episode, faith in television weather forecasting as an explanatory system is placed against the efficacy of Jangala's rainmaking skills (demonstrated by the storm that engulfs the film shoot), and the mysterious power of his Ford V8 painted with the Rain Dreaming.

Big themes weave through these apparently simple storylines: physical and cultural survival; the centrality of country; the ongoing importance of ceremonial life and law; the contemporary economy of hunting and collecting (for both food and spare parts); the everyday proximity of people who have gone before; the material deprivation of *yapa* life and the physical desolation of the Yuendumu township. Here viewers are offered glimpses into other, radically different Australian lives and other, radically divergent perspectives on a familiar technology. While the landscape is instantly familiar as the iconic 'dead heart' of the continent, most Australians choose to remain oblivious to the realities of Aboriginal life in the Centre – except perhaps vaguely, as an intractable 'problem'.

Instead of confirming this view, the episodes provide a humorous vision of life under settler colonialism from a Warlpiri perspective. Viewers are thrown straight into the lives of these men, with little set-up, as if we already know them. By showing the 'indigenisation' of imported technologies – such as cars, radio, rock music and filmmaking – and the cultures that surround them, the series demands a re-imagining of what is too often called 'one Australia'. Intercultural difference and fluidity is highlighted in the series and manifested in the multilingual dialogue, which flows between three languages – Warlpiri (the predominant language, minimally subtitled into English), Kriol, and some English. A further dimension is added with the distinctive sign language used throughout the series without explanation or comment.

Something of the random and inexplicable logic of colonialism is conveyed in the ways the men respond to the many things that go wrong with their cars. The endless succession of mechanical setbacks are ordinary, everyday occurrences and there seems little use in trying to anticipate (or even be annoyed by) them. Non-Aboriginal people – be they police, prison guards, the service station owners who control the fuel that the bush mechanics need, car salesmen, or tourists encountered on the road – are a peripheral presence. A Warlpiri view is the only one that matters here, and outsiders are treated with varying degrees of wariness, occasionally expressed with sardonic and understated shorthand ('Cheeky police here in Broome'), but that is all.

There is a strong historical dimension to the series, which places contemporary events within a longer tradition of Warlpiri car culture. *Yapa* first encountered trucks as a mysterious and frightening intrusion: supernatural monsters, perhaps. Elder Jack Jakamarra Ross recounts memories of those times directly to the camera, while his young self is played on screen by Francis Jupurrurla Kelly. Kelly's comic lightness of touch brings life to Warlpiri stories of first contact at the same time as it contains more than a hint of irony at white fantasies of the 'noble savage'. The trucks were creatures whose tracks were unfamiliar and threatening, Jakamarra Ross declares, all the more so because they did not appear to shit. He tells us that the Warlpiri knew then that they would need to be really careful of them. Before too long, trucks were understood to be instruments of Europeans' power – power to invade their country, power to remove Warlpiri from their country by taking them to prison at Jay Creek for spearing cattle, and eventually to the settlement of Yuendumu. There it was intended that their mobility would be halted, though as the series clearly demonstrates, that aim has never been realised.

A derelict truck, now almost hidden in grass and scrub, remains alive to these storytellers as a witness to that past. Its rusting body has become a potent container for memories of terror and dispossession, most notoriously the Coniston Massacre of 1928. For months, scores of Warlpiri men, women and children were hunted across large tracts of county and shot by white settlers in reprisal raids that have never been properly investigated, nor the perpetrators held to account.[1] The Coniston station truck, we are told with chilling understatement, was one that Aboriginal people definitely did not choose to ride in. Recollections of violent dispossession, as well as subsequent pleasurable memories of the cars that *yapa* themselves have owned and loved, inhere in the wrecked cars and trucks scattered in the township and throughout the bush. In practical terms they are repositories of spare parts, but they also have become biographical markers for individuals and the community, just as 'natural' features of the landscape like rocks, trees and waterholes also have their stories.

While memories of the humour, indignity and anguish associated with a technology of invasion opens the *Bush Mechanics* series, the stories soon turn to the ways that Warlpiri themselves have taken to automobiles. While Europeans first brought motor vehicles into the Central Desert, now these Warlpiri men leave the community to bring them in for their own purposes and according to their own values. Cars and the things that *yapa* do with them have become saturated with meanings and embedded in practices that reflect the possibilities and constraints of life in Yuendumu. Perhaps most notably, the distinct car culture portrayed in *Bush Mechanics* is a contemporary expression of an earlier subsistence economy that continues to be strong in the present.

Crocodile, the dapper advocate of humpy living in the fourth episode, wonderfully demonstrates the ongoing importance of bush food in contemporary Warlpiri life. As he digs a kangaroo out of the ashes, Crocodile proudly declares that he is able to provide his family with a big feed without a supermarket. Automobiles are similarly placed, by choice and by necessity, within a local subsistence economy that continues to thrive alongside the cash economy. By incorporating cars within that system, the bush mechanics have devised ways around the material deprivations that characterise their lives in Yuendumu. They have created their own ways of being men with wheels, based on an impressive disregard for the orthodoxies of individual car ownership, the economics of the car market, and the professionalisation of automobile repair. In so presenting the bush mechanics as mobile makers – foragers of mechanical parts and disseminators of alternative solutions – the series offers upbeat parables of the men's self-determined survival within settler colonialism.

Bush Mechanics serves to remind us that the mechanical creativity portrayed in the series is not a recent talent for Aboriginal people, nor is it confined to the Central Desert. Indigenous people across the continent have long valued automobiles and found ingenious ways to bring them into their personal and collective lives. There are records from as early as 1917 of an East Arnhem Land man's enthusiastic response to cars. He was identified as 'Davy Fanchala' by Francis Birtles, the restless overland 'adventurer' who spent decades traversing Aboriginal land on his bicycles and cars. Birtles recorded travelling from Burketown to the Calvert River with the help of Davy, whom he called his 'primitive brother'. Davy 'soon learnt what a motorcar could do and where it could go', Birtles wrote in one of his regular newspaper articles.[2] Together they dragged his Ford through flooded creek beds and along bush tracks that were not designed for cars. They overhauled the engine, straightened bent axles and repaired a damaged radiator. Davy's good tricks call to mind the Yuendumu bush mechanics' humour. Davy loved to startle his friends, Birtles recorded, by getting them to 'join hands, and then get hold of the spark plug wire and switch on the dry cell electric current'.[3] Davy composed songs about the experience of driving, in which the 'swift-swaying tones of his melodious voice suggested the rhythm and movement of a fast-running motorcar'. When Davy returned home, his songs were turned into a 'motorcar corroboree', Birtles wrote, and staged for countrymen who came hundreds of miles to see the 'marvellous performance'.[4]

Like Davy, Aboriginal people have seized opportunities to become auto-mobile wherever and however they were able. From the 1920s, some men learned to repair machinery at blacksmith shops in mission stations, and their skill as innovative bush mechanics became legendary.[5] At a time when state governments were confiscating Aboriginal people's wages, some managed to maintain control over the money they earned from land clearing, fencing, boring wells, shearing, mining, timber-cutting, stock work, rabbit-trapping and dingo-scalping. Some bought used cars and trucks, to the consternation of missionaries and the envy of their white neighbours.[6] Even as early as 1915, newspapers in the Northern Rivers district were reporting that Mr Combo, of the Aboriginal 'colony' near Ballina, dared to own both a motorcar and a motorcycle with sidecar.[7] Car ownership was easiest in areas where there were labour shortages, such as in rural South Australia, and where unions insisted on equal pay for Aboriginal workers. Non-Indigenous family or friends sometimes helped out by posing as the owners of Aboriginal people's motor vehicles.[8] Families sometimes clubbed together to buy and maintain a vehicle and many photograph collections contain snapshots of family and friends dressed in their finest clothes, posing in front of their prized vehicles.

White travellers on the Nullarbor Plain reported being startled to encounter Aboriginal people driving cars in the 1920s, well before non-Indigenous workers were generally able to afford them. A missionary from the Bush Church Aid Society, for example, stopped and photographed two Aboriginal men who owned a car in 1925. He included that photograph in his account of the journey in a Sydney newspaper.[9] Further west, near Colona Station in 1928, members

The inevitable Ford, near Collona Station, with two aboriginal passengers.

Melbourne Argus, 26 May 1928. National Library of Australia.

TWO ABORIGINALS WHO OWN A MOTOR CAR. PHOTOGRAPHED ON THE EDGE OF THE NULLABOR PLAINS.

Above: *Sydney Morning Herald*, 11 July 1925. State Library of New South Wales.

Right: 'Dick Davey's Family and Car' (original caption), 1927 Koonibba Lutheran Mission, South Australia. National Library of Australia Album 1170.

of the MacRobertson truck expedition, which was lumbering around the continent promoting their sponsor's sweets, were astonished to see an Aboriginal man and woman driving an old Ford car. They stopped to take a photograph and presented them with a box of MacRobertson's Maxmints.[10] A photograph taken by Herbert Basedow, on one of his expeditions through South Australia and the Northern Territory in the 1920s, showed a stylish man, whose name he did not record, relaxed and at home behind the wheel of a Ford car.

Reports sent back by travellers passing through Madura Station at that time similarly reveal that Aboriginal workers there were enthusiastic and creative bush mechanics, skilled at cobbling together vehicles out of the materials at hand. One traveller wrote of the 'many and ingenious' vehicles that Aboriginal men constructed. He described their cars as 'masses of wire and string holding the boards and mallee sticks together'. Some owned 'high body cars obviously of the model produced in or about the year 1914', which are 'packed with the family, dogs and camping and hunting outfits', he wrote. 'Fitted with tyres and tubes that have far outlived their usefulness, there is constant trouble and the proud owners find themselves little better off than their brothers who are content with the ship of the desert and a cart'.[11]

Aboriginal people at the Lutheran mission station at Koonibba owned cars in the 1920s, even before the missionaries themselves were able to afford one. The superintendent, in his annual report to the Chief Protector of Aborigines, wrote with alarm in 1926 that a number of residents owned motorcars. He declared that unscrupulous settlers were hiring Aboriginal people to clear the mallee scrub from their newly established farms, and paying them with 'a conglomeration of scrapped tin and boards'.[12]

It is clear, however, that Koonibba men were not the gullible, mechanical innocents the mission superintendent liked to imagine. Both Dick Davey and Yarrie Miller were prominent residents of the mission at that time, football heroes and gun shearers. To this day they are remembered as skilled bush mechanics and proud car owners. In 1927 the Davey family, dressed in their best for a special event, perhaps their baby's christening, posed for a photograph with their immaculate Ford. The image was carefully preserved in the mission's photograph album, as a symbol of the good work they were doing. The album was recently deposited in the National Library of Australia.

Eighty years later in 2010, a former missionary at Koonibba wrote about the famous story of Yarrie Miller, Dick Davey's friend, and his remarkable drive home from a shearing season in the early 1920s. 'On the way home,' as the missionary told it, 'a piece broke out of the cylinder block, leaving a hole in the cylinder wall. Nevertheless, Yarrie arrived safely back in Koonibba, with the help of his brother Bill, who lay on the running board and mudguard at the side of the engine feeding oil into the hole in the cylinder wall, so that the motor did not run dry and seize.' Yarrie Miller's car was still a 'notable landmark' in the scrub at Koonibba forty years later, the missionary recalled. The wreck was valued as a powerful reminder of Nunga men's pride, independence, skill and mechanical creativity.[13]

Still in South Australia but further northeast in the Flinders Ranges, Adnyamathanha men at the United Aborigines Mission camp at Nepabunna had long been important providers of transport services. They were noted as owners of camels (Mount Serle had been a government camel depot) as well as valuable donkey teams, which they used to take pastoralists' wool clip to the railway. During the 1920s, newspapers reported that there were some eleven cars and trucks at the camp – that is, until unemployment, drought and hardship during the Great Depression reduced the residents to starvation and poverty.[14] Their motor vehicles, however, continued to be useful and in *yura* (as Adnyamathanha call themselves) hands they were creatively reimagined in ways that car manufacturers had not envisioned. When fuel and spare parts were hard to find or too expensive, *yura* stayed mobile by harnessing their cars to donkeys. For decades, those Nepabunna cars pulled by animal power were known as 'bun carts'. The anthropologist Charles Mountford photographed Tim Coulthart in 1935, leading a bun cart carrying Rosie Brady and a young Frank Driver.[15] Twenty years later they were still in use, and one caught the attention of author Frank Clune as he passed through the Flinders Ranges in 1953. For decades, *yura* used bun carts for transport work, cultural business, holidays, collecting rations, carting water to camp, children's play, and for carrying bodies for burial.[16]

During the interwar years, Aboriginal people were increasingly able to own automobiles. Motor vehicles were highly valued in Aboriginal communities but they were used and owned somewhat differently than by their white neighbours. Distinctive Aboriginal mobilities were shaped by cultural differences and material impoverishment, as well as the need for protection from the corrosive impacts of white supremacy. Cars and trucks made it easier for Aboriginal people to earn money; to stay in touch with scattered kin; to organise politically; and to keep one step ahead of the welfare. They provided places to sleep in; dignified conveyance for funerals, confinements and weddings; transport for picnics, holidays and shopping trips; rapid aid for medical emergencies; and a buffer against racism on public transport.[17] When better times returned to the Flinders Ranges in the 1940s and 1950s, *yura* again began to buy cars and trucks, occasionally even new ones, to the renewed astonishment of their white neighbours. Journalists in Broken Hill, for example, wrote stories about the 'strange' sight of Aboriginal

A Nepabunna 'bun cart', 1935. Photographer: Charles P. Mountford. State Library of South Australia PRG 1218/34/834D.

families driving the long distance from Nepabunna in their trucks to take a holiday, to visit friends, and do their Christmas shopping.[18]

Throughout the twentieth century, Aboriginal people increasingly used motor vehicles to adapt to settlers' economic and social arrangements, but also to strengthen and extend their pre-colonial cultures. Motor vehicles became indispensable for hunting and gathering the foods that white incursions into their country had made increasingly scarce; for maintaining family ties across distances; for moving away from reserves and missions to establish smaller outstations on homelands; for reviving and revitalising ceremonies that had gone into decline; and for producing the wealth of artwork now held in collections around the world.

The *Bush Mechanics* series, then, is just one example of longstanding Aboriginal connections to motor vehicles, which took a very different form from what we might loosely call 'mainstream' settler Australia. For while Aboriginal and non-Aboriginal people moved together, often on the same roads and tracks, they did not always move in exactly the same ways. Aboriginal people built their own automobile cultures in response to dispossession and displacement from their country, and so remained highly mobile. They embraced automobility with enthusiasm whenever and however they could, not by simply imitating mainstream car cultures, but by re-crafting automobility according to their own cultural imperatives and the material possibilities open to them. Cars and trucks, no matter how battered or worn, were prized objects that were amenable to uses manufacturers had not envisaged. They were wide open to material modification and cried out for cultural re-inscription. Through them, as the *Bush Mechanics* series so wonderfully shows, Aboriginal people enacted new stories, devised new kinds of pleasures, created new practices, accommodations, resistances, and collaborations with the colonial system that redefined their lives throughout the twentieth century.

Now in the twenty-first century, users are discouraged – and even penalised – for repairing, servicing or modifying their cars. Tinkering of any kind, let alone the radical surgery performed by the bush mechanics, is discouraged by the growing complexity of motor cars, the cost of specialist tools, the risk of having a warranty voided, as well as the government regulations that set safety standards. Cars are increasingly sealed objects, given by experts and designed with no room for consumers to move – except for the kind of authorised movement that has been built into them. That flattening of possibility, no matter how scientific and rational, no matter how much it seems to inhere in the natural evolution of 'generations' of constantly improving cars, and no matter how much it might be welcomed as 'progress', embodies political choices about how our lives should be lived, how we may act, and who can take responsibility for what. The *Bush Mechanics* series provides a timely reminder that the 'given-ness' of the technological worlds we inhabit should always be called into question.

In the third episode, Jack Jakamarra Ross, speaking from a radically different context from that of mainstream car cultures, lovingly touches what might appear to be a useless engine

'Aborigine with his T Model Ford, "Tuncoona" Station – Bourke, NSW' (original caption), c. 1934. State Library of New South Wales 391737.

block inappropriately dumped in the bush. Translated into subtitles, he quietly declares its value to his community: 'This motor grew us up. Now it is lying here like a witness looking after us.' It is a startling statement, a key to how we might understand the importance of the *Bush Mechanics* series. Jakamarra Ross's declaration points to the heart of how technologies are taken up into social life. They change people's lives, but are also are shaped by people into something new – something very different from what was first encountered. *Bush Mechanics* shows, in other words, how cars continue to be manufactured by consumers after they leave the factory floor.

This series articulates one particular Warlpiri view, which takes tremendous pleasure in unauthorised experimentation and change. In declaring wrecked motorcars a kindly witness to past times, Jakamarra Ross invites his non-Aboriginal audience to suspend their automobile habits and challenges us to ask fresh questions. Perhaps we should all ask questions of how cars have differently grown us up. What kinds of knowledges and distribution of powers have cars made seem natural, and what desires or fears do they bear witness to? For non-Aboriginal Australians, might our privileges also be our loss? Are there more inclusive or democratic terms through which we might together re-imagine and re-draft the power to move? Certainly, we urgently need to find ways that will open up the possibilities of alliances – and justice – across racial difference. Clearly, we also need to devise ways to make movement on this continent sustainable if we are to live here for a further 65,000 years. There is much to learn – and much to be done.

ACKNOWLEDGEMENTS

This chapter is based on an earlier article I published in *Australian Humanities Review*: see 'Still Moving: Bush mechanics in the Central Desert', *Australian Humanities Review* volume 25, March 2002.

See also my 'Mobile Encounters: Bicycles, cars and Australian settler colonialism', *History Australia* volume 12, issue 1 (2015), pp. 165–185; and '"Australia – Drive It Like You Stole It": Automobility as a medium of communication in settler colonial Australia', *Mobilities* (October, 2017).

The research for this chapter was supported by an Australian Research Council Discovery Project [grant number DP110101875], 'Mobile Modernities: Around-Australia automobile journeys, 1900–1955'. My thanks go to Dr Jeannine Baker for her research help on this project.

Georgine Clarsen is an Associate Professor in History and Politics in the Faculty of Law, Humanities and the Arts at the University of Wollongong. She is also an Associate Editor of the new journal, Mobilities: Interdisciplinary Journal of Mobility Studies *and has worked as a consultant and talking head on the acclaimed documentary series* Wide Open Road: The Story of Cars in Australia *(Dir. Paul Clarke, Bombora Films) which screened on ABC TV in 2012.*

NOTES

1 *Coniston*, dir. David Batty and Francis Jupurrurla Kelly, PAW Media and Rebel Films, 2012, http://coniston.pawmedia.com.au/home.

2 *Mirror*, 13 October 1917, p. 11.

3 *Mirror*, 20 October 1917, p. 11.

4 *Mirror*, 13 October 1917, p. 11.

5 Peggy Brock, *Outback Ghettos: Aborigines, institutionalisation and survival*, Melbourne: Cambridge University Press, 1993, pp. 164–165; *Adelaide News*, 21 November 1930, p. 16.

6 *Western Champion*, 19 March 1925, p. 9; *Western Mail*, 27 October 1932, p. 7; *Courier Mail*, 6 April 1938, p. 11.

7 *Richmond River Herald and Northern Districts Advertiser*, 8 October 1915, p. 4.

8 Peter Gifford, *Black and White and In Between: Arthur Dimer and the Nullarbor*, Perth: Hesperian Press, 2002, pp. 89–90.

9 *Sydney Morning Herald*, 11 July 1925, p. 9.

10 Georgine Clarsen, 'The 1928 MacRobertson Around-Australia Expedition: Colonial adventuring in the twentieth century,' in *Expedition into Empire: Exploratory journeys and the making of the modern world*, edited by Martin Thomas, London: Routledge, 2014, pp. 194–213.

11 *Adelaide Advertiser*, 9 August 1930, p. 7.

12 *Adelaide Advertiser*, 18 November 1926, p. 16.

13 C.V. Eckermann, *Koonibba: The mission and the Nunga people*, Clarence Gardens, SA: Elizabeth Buck, 2010, pp. 146–147.

14 *Adelaide News*, 21 November 1930, p. 16.

15 Individuals in the photograph were identified by Vera Austin (nee Coulthart) in an interview with local historian John Mannion in 2013.

16 Brock, *Outback Ghettos*, p. 164.

17 Ursula Frederick, 'Roadworks: Automobility and belonging in Aboriginal art', *Humanities Research*, vol.17, issue 2, 2011, pp. 81–107; Denis Byrne, 'Difference', in *The Oxford Handbook of the Archaeology of the Contemporary World*, edited by Paul Graves-Brown and Rodney Harrison, Oxford: Oxford University Press, 2013, p. 295.

18 *Barrier Miner*, 22 December 1952, p. 2.

'WE HAVE ALWAYS MOVED AROUND'

Backstories on Warlpiri mobility and media

Melinda Hinkson

Melinda Hinkson traces the recent past of Warlpiri people and the settlement of Yuendumu, focusing on the rich history of media production.

Jampijinpa, a senior man aged in his sixties, dismisses with a wave of his hand my suggestion that the time of settlement was the time when Warlpiri sat down in one place. 'We have always moved around'. Indeed, an impetus to move emerges as a continuous theme across Warlpiri culture and history – from the journeying trajectories of ancestral heroes; to old people's memories of walking all over the desert before the coming of *kardiya*, white man; to the forced migration of settlement time; to the droving and other kinds of work that took some people thousands of kilometres across the continent; to the more recent travel for shopping, to exhibit art, to visit family and for all manner of reasons via motor vehicle and aeroplane. *Bush Mechanics* hints at a much larger Warlpiri story about the integral place of journeying as a creative response to momentous change. *Bush Mechanics* was made by and associated with people who live at Yuendumu; but the people of that place do not sit still and the place itself continues to transform in tandem with the changing world around it.

Yuendumu was first established as a ration depot in 1946 and since then has at different times been known as a 'settlement', a 'community' and most recently a 'town', even a 'growth town'. Each of these descriptions carries with it a weight of government policy imagination and a set of aspirations for how and where Yuendumu's residents might be expected to live. Across that time Warlpiri have themselves responded creatively and energetically to each new set of challenges governments have posed. Today Yuendumu is home to a fluctuating population of around 1000 people. The majority of those people are Warlpiri, or *yapa*, as they refer to themselves.

Warlpiri are owners and custodians of a vast area of the Tanami Desert, one million hectares of predominantly open spinifex plain traversed by sand dunes, spectacular mountain ranges and rocky outcrops. The desert comprises a rich patchwork of intersecting countries and Dreamings. *Jukurrpa* is the Warlpiri term for Dreamings, a body of law that lays the foundations for the world order. It tells of the world-making journeys, contests, passions and tragedies that beset ancestral heroes. Across the desert these ancestors left traces of their actions in the form of *kuruwarri* – footprints, bodily impressions, resting places – as sites that hold the sedimented

essence of ancestral power. The mythic stories of the ancestors in turn supported the journeying life of desert-dwelling hunter-gatherers, presenting people with a moral and social order through which to relate to country, but also, in practical terms, showing them where to find water, how to ensure plentiful supplies of bush foods, and what to be wary of. Old people who grew up in the desert characteristically sing *Jukurrpa* songs as they journey by car to locate places not visited for some time.

Wally Japaljarri: *The Warna* [snake] *at Kaltarrangu*, Hooker Creek 1953–1954. Drawing #122, Meggitt Collection, Australian Institute of Aboriginal and Torres Strait Islander Studies.

'BULLOCKS AND BLACKS DON'T GO TOGETHER'

From the 1920s life in the desert changed irrevocably. Growing numbers of prospectors were being drawn to Central Australia from the southern states, chasing the promise of gold. Pastoral leases were issued across large tracts of territory, prospecting leases were pegged out and Aboriginal people increasingly found themselves in competition for access to their precious hunting grounds and water sources. Prolonged drought further exacerbated tensions between Aboriginal people and settlers. Evidence suggests that by 1928 Warlpiri and their Anmatjere neighbours were starving.[1] Through this period there were incidents of Aboriginal people killing stock, perhaps in retaliation for being driven away from water sources as well as to restore 'self-confidence and prestige'.[2] There were well-publicised incidents of settlers being acquitted over brutal deaths of Aboriginal persons across inland and north Australia. In Warlpiri country, several men were arrested at the Granites in association with the death by spearing of a miner. They were taken to Darwin to face court but later discharged for lack of evidence. With their case dismissed these men were released on the outskirts of Darwin. Their families never heard from them again. Others were seized by miners on the goldfields and simply vanished.[3]

Expedition members panning off prospects near Crown Creek, close to Coniston Station, Northern Territory, 1928.
Photographer: Michael Terry. National Library of Australia PIC/8847/9/91.

All these tensions came to a head for Warlpiri at Yurrkuru on the Lander River in August 1928, when a Ngalia (Southern Warlpiri) man named Kamalyarrpa Japanangka, also known as Bullfrog, killed dingo trapper Frederick Brooks. This killing was in retaliation for Brooks's failure to act honourably in an exchange of services provided by Japanangka's wife, Marungali Napurrurla. In the weeks that followed, Constable George Murray led a party of local settlers and Aboriginal trackers on reprisal killings in the most recent recorded mass killing of Aboriginal people in Australia, which would later become known as the Coniston Massacre. Murray's party hunted down, shot and killed men, women and children. The actual number of people who perished in these raids is unknown. A government inquiry found that thirty-one people died. Warlpiri estimate a number closer to 100. In the early 1990s Petronella Vaarzon-Morel recorded the memories of women who as children had survived these terrifying raids. Rosie Nungarrayi recalled:

The policemen came travelling this way, north, after the shooting time. They came from Pijaraparnta. That's were they came upon a lot of people and slaughtered them, our relatives. This happened when I was a little girl.

After that, the policeman came to Liirlpari [Whitestone]. Again the policemen travelled west to Patirlirri [Rabbit Bore] looking for people. Again they killed a lot more of our people . . . For a whole day they went around shooting at people. They shot them just like bullocks. They shot the young men coming out from the bush camp where they'd been initiated. People were shot digging for rabbits in our country, Muranjayi. They were getting yakajirri berries, yams and wanakiji tomatoes. Those policemen shot them for nothing. Again they killed a lot of men there. No one breathed. All were dead . . .[4]

A series of inquiries exonerated the men involved in the Coniston killings. In the submissions made in court much is revealed of wider settler attitudes to Aboriginal people in this period. C.H. Noblett, who held joint positions of Chief Protector of Aborigines and Chief of Police, made clear he saw the killings as unfortunate but inevitable. They were acts undertaken in support of a greater good: colonial nation-making:

I deplore the killing of the natives as much as anyone but, at times, it cannot be avoided and the same thing has happened in the settling of all new countries. Lessons must be taught to people who murder others . . . settlers are very fair to the natives but stock and natives do not and will not thrive together . . . If this industry is to be settled with a healthy white population, we must give the pioneers every protection both for themselves and their stock otherwise the country must be left to the natives who have not the slightest idea of development in any shape or form.[5]

At the heart of Warlpiri country was Pikilyi, an estate containing a network of natural springs and water holes, with lush surrounding country that was replenished by these waters and

Following spread: Brooks Soak near Coniston Station, Northern Territory, 1928. Photographer: Michael Terry. National Library of Australia PIC/8847/11/100.

sustained rich hunting grounds. Pikilyi was literally an oasis in the desert and the place at which Warlpiri congregated in large numbers for ceremonial gatherings. In 1932 returned serviceman William Braitling was granted a pastoral lease over Pikilyi and 2,500 square miles surrounding it. He established a homestead he named Mount Doreen, in honour of his wife. In the same period as Braitling commenced stocking the property he discovered rich deposits of wolfram, a mineral widely used in the production of ammunition that was in high demand in the lead-up to the Second World War. He set up a ration depot close to the Mount Doreen homestead and lured Warlpiri to work in the wolfram mines in return for rations. Braitling was infamous for his harsh treatment of people camped in the vicinity of his homestead and a series of reports by government officials and missionaries detailed incidents of malnutrition, illness, brutality and suspected improper dealings with young women.

'Tucker time for wolfram gatherers, Mount Doreen, Northern Territory' (original caption), 1952. Photographer: Arthur Groom. National Library of Australia PIC/5879/118.

Harry Jakamarra Nelson told of a team of men pushing a wheelbarrow of wolfram from the mine at Mount Singleton to the depot at Luurnpa-kurlangu, a distance of more than seventy kilometres. Braitling was charged but found not guilty of causing grievous bodily harm to Jimija Jungarrayi, a man who would go on to be a prominent leader, and who Braitling had helped raise from childhood.[6] In 1940 missionary Laurie Reece observed Warlpiri working in dangerous conditions in the mine and 'wondered what the reaction of a body such as the Anti-Slavery League in London would be'.[7]

As Warlpiri became increasingly dependent upon western foodstuffs such as flour, sugar, tea, and tobacco they moved between the places where these supplies were on offer – the Granites goldfields, Haasts Bluff ration depot, and Mount Doreen – often leaving one set of hostile conditions for another that was only marginally better. Their movements were observed and reported by patrol officers, missionaries and other dedicated advocates such as Olive Pink, who campaigned for the establishment of reserves and settlements to protect Aboriginal people from the exploitation that had become widespread.[8]

SETTLEMENT TIME

A ration depot was established by the government at Yuendumu, on the southern edge of the Tanami Desert, in 1946. It soon attracted a large number of people. Baptist missionaries in conjunction with the Commonwealth government set about building infrastructure for a permanent settlement. Warlpiri were enlisted to grade roads, clear an airstrip and construct buildings including a hospital, dining hall, church and mission house. In time they would be put to work as bakers, cooks, cleaners and nurses. As more people moved in from the desert they

Left: Aboriginal man carrying a steer's head, Mount Doreen, Northern Territory, 1946. Photographer: Axel Poignant. National Library of Australia PIC/11868/62. Courtesy Roslyn Poignant.

Right: Yujuku shelter with Kingstrand huts in the background, Hooker Creek, c. 1953–1954. Photograph: Mervyn Meggitt, Meggitt Collection, AIATSIS, N390.131.

established semi-permanent camps to the north, west and east of the settlement, oriented in the direction of the countries from where they had come. In the settlement camps people continued to erect shelters similar to those they assembled in the desert, but gradually incorporated corrugated iron and other building materials found to be useful in repelling rain and wind.

Early on Yuendumu was beset with challenges of over-crowding and insufficient water supply. Moves were made to set up a second settlement in the northern Tanami Desert, initially at Catfish. When the water supply was found to be inadequate the settlement was moved to Hooker Creek, some 600 kilometres north of Yuendumu. A small team of men were enlisted to grade the road and begin construction of the new settlement. In early 1952, 100 people residing at Yuendumu were boarded onto two government trucks and relocated to Hooker Creek. Decades later, people recall this upheaval with a sad sense of resignation. How could they say no to the white man? Others were reportedly enthusiastic at the prospect of new adventures in unknown country and relieved to be leaving a troubled and over-crowded place. The site to which they were taken however was dangerously ill-equipped to deal with the arrival of such a large number of people. It was also foreign country, land owned by the Gurinji. In the months that followed a number of small family groups who were racked by unhappiness, missing kin and country to the south, walked the hundreds of kilometres back through the desert to Yuendumu.

YUENDUMU IN THE WORLD

For twenty-five years, Yuendumu's residents were presided over by Baptist missionary Tom Fleming alongside a series of government superintendents. Fleming held regular film nights and many people got their first views of the world beyond Central Australia from these screenings. Yuendumu was both the largest settlement in the region and also relatively accessible, located just 300 kilometres from Alice Springs. It became a popular destination for politicians, bureaucrats, journalists, researchers and film crews enquiring into all aspects of Warlpiri life and culture. Indeed scientific and anthropological expeditions had visited the wider region since as early as the 1880s. From the 1950s, government-sponsored filmmakers produced a series of short films recording aspects of classic Warlpiri cultural production, as well as propagandist-style documentaries reporting on the success of the assimilation program in transforming desert-dwelling hunter-gatherers into a sedentary and civilised population.

Neville Japangardi Poulson's picture of the moon landing, Yuendumu, c. 1969. Slide courtesy David Tunley.

Through the 1950s and 1960s Warlpiri moved out of the camps to occupy the first generation of settlement housing. Children attended school where they were exposed to new ideas and strict authority. Each morning they were washed and clothed in special-purpose school clothes. They were prohibited from speaking Warlpiri in class and forbidden from leaving the school grounds during school hours. Once released, children retreated to the bush at every opportunity, sometimes on the backs of donkeys. From the late 1960s, children's experience of the wider world expanded further. They were taken on school camps and excursions beyond the Northern Territory. They were encouraged to imagine and depict events occurring in the world at large in drawing and painting classes.

Those who completed school and went on to pursue further study travelled to teachers' colleges and universities in Darwin, Adelaide and Melbourne. They returned home with new insights into the places they had visited and the diverse ways of living they had observed in metropolitan centres. Significantly, in this period of assimilation people were still able to enjoy time in the desert. During the long summer breaks, as was usual practice across Aboriginal settlements, Yuendumu's Warlpiri residents were relocated to a bush camp. The summer camp was a cost-saving measure for the settlement, but paradoxically the wholesale relocation of the community to the bush also allowed elders to practise customary forms of authority largely unimpeded by the interventions of missionaries or superintendents. Initiation ceremonies were held and days were spent hunting and gathering bush foods.

Another series of changes was ushered in from the 1970s. National momentum toward the recognition of Aboriginal land rights encouraged Aboriginal people of the Northern Territory to pursue their aspirations to establish small living areas on their ancestral estates. Simultaneously, a new policy of self-determination drew Aboriginal people into the work of transforming the 'settlement' into a 'community'. This process was to be significantly informed by Aboriginal imperatives, priorities and cultural practices. While in the 1950s children had been banned from speaking Warlpiri in school, by the 1970s language revival was introduced by way of a bilingual education program. A new Bilingual Resources Development Unit staffed by a teacher-linguist and trainee Warlpiri teachers produced dozens of school readers in Warlpiri language to enable the reorientation of the curriculum. Senior men were invited to paint the doors of the school classroom with *Jukurrpa* designs.[9] In the place of the superintendent, a community government council was established with elected Warlpiri representatives.

From the 1980s a suite of new community organisations was formed to support the flourishing of cultural production and new enterprise. An artist's association, crafts association, women's centre, housing association, cattle company, small mining exploration company, outstation resource centre, social club and community store were all established at Yuendumu, with Warlpiri boards of management. Many of these new ventures received government grants and government-funded motor vehicles. For the first time since being forced into a relatively sedentary life on government settlements Warlpiri had the ability to venture back into the desert.

Access to vehicles gave people unprecedented freedom to travel spontaneously for hunting trips, to participate in ceremonies, to visit kin in neighbouring towns, to travel to regional centres for shopping, and to pursue adventures further afield.

Perhaps paradoxically, the 'return to country' movement, as the movement to create outstations became known,[10] coincided with another set of developments that would give remote living Aboriginal people unprecedented exposure to the world at large. In 1986 the Australian government launched AUSSAT, the first generation of Australian-owned satellites. This satellite would beam national radio and television transmissions across inland Australia for the first time.

PICTURING THE WORLD YAPA WAY

Yuendumu's residents had been experimenting with their own video production and local news broadcasts since the early 1980s. Video equipment had initially been introduced in an adult education program and was then expanded with the support of a research project funded by the then Australian Institute for Aboriginal Studies.[11] In the lead-up to the launch of the satellite, people from Yuendumu campaigned in conjunction with other Aboriginal communities including the Alice Springs-based Central Australian Aboriginal Media Association. They gave voice to remote Aboriginal people's concerns, especially the concerns of older people, regarding the threats posed by the introduction of national television.

What were they concerned about? As a form of 'mass' media, television's democratic reach functions in a way that was seen to be thoroughly at odds with the customary control and exchange relations of Aboriginal society, where kinship, gender, generation and place-based relationships determine who has the authority to know, to speak, or reveal images and stories. A person inherits rights and responsibilities to place-based knowledge through their father's and mother's lines. These reciprocal kin-based ways of practising knowledge are acquired by a person and enacted socially in ceremonial gatherings, where the performance of a particular song, for example, can only occur with the presence, participation and direction of kurdungurlu ritual managers. The same principles and concerns govern any activity associated with ancestral places, and more broadly pervade day-to-day social interactions and especially any kind of public discussion. In relation to the launch of television, senior people were especially worried about the transgressions of customary authority that would occur if archival films of restricted men's ceremonies were broadcast into their communities.

The government responded to these appeals for sensitivity by introducing a special Broadcasting for Remote Aboriginal Communities Scheme (BRACS) in tandem with the launch of the satellite. The BRACS program delivered specially designed re-transmission equipment that enabled community residents to receive and retransmit two radio channels and two television stations. In deference to concerns about the need to have some control over the circulation of material coming into communities, the equipment included an interruption

George Marshall at Frederick Brooks' grave, from *Coniston Story*. Pintubi Anmatjere Warlpiri Media.

switch that allowed local residents to insert their own content in place of any programs deemed unsuitable for retransmission.

As Warlpiri took up video equipment for their own purposes, they did so with distinctive ways of relating to each other and seeing the world. Researcher Eric Michaels, working closely with Francis Jupurrurla Kelly, described the aesthetics of early Warlpiri media, tracking the way image-making was shaped by the imperatives of kinship and ways of relating to country informed by the ancestral order. In the making of *Coniston Story*, a video that recounts the events associated with the 1928 killings, Michaels described a set of complex social requirements that had to be negotiated. While the video appears to be a relatively straightforward sequence of country shots and interview, behind the scenes there were more than twenty people who travelled to the filming location in several cars. They were *kirda* (owners) and *kurdungurlu* (managers) for the country in question and their presence was required to authorise the story being told by old Japangardi, the son of Bullfrog, as *junga*, 'proper', or true.[12] Kin-based authority structured relationships on both sides of the camera, determining for example, who could interview whom. Sequences of long pans of the surrounding country coming in and out of focus, which might have been dismissed as naïve filmmaking, were in fact deliberate techniques, following the movements of ancestors over the hills and into the foreground.

Reflecting on the world-changing implications of introducing broadcast media into Warlpiri communities, Michaels wrote of what he saw as the clash of two cultural systems diametrically opposed in their logics:

The bias of mass broadcasting is concentration and unification; the bias of Aboriginal culture is diversity and autonomy. Electronic media are everywhere; Aboriginal culture is local and land-based.[13]

To ensure Warlpiri were able to make media that supported their cultural imperatives, Michaels promoted what he dubbed 'a cultural future':

... an agenda for cultural maintenance which not only assumes some privileged authority for traditional modes of cultural production, but argues also that the political survival of indigenous people is dependent upon their capacity to continue reproducing these forms.[14]

Different generational attitudes to the introduction of television were evident. Senior Warlpiri man Darby Jampijinpa Ross made a stirring address to a Department of Communications inquiry in 1984, telling government bureaucrats, 'We got our land back to stop whitefellas chasing us with things like satellites'.[15] At the same time younger people were enthusiastically taking up video cameras and immersing themselves in the new worlds offered by television.

The Warlpiri Media Association was incorporated in 1986. The charter adopted by the organisation gave it authority to act on behalf of community interests in relation to all aspects of media activity, including liaising with film crews and journalists to ensure any filming activities in and around the community were conducted appropriately and respectfully. Early media activity by Warlpiri media staff both exemplified and recorded the flavour of social activity in this dynamic period – video records were made of culturally engaged school-based activities, trips to country by people visiting places for the first time in many years, old people reminiscing, and meetings between residents and visiting bureaucrats and politicians about all manner of community development issues. In this way, video cameras were witness to a period of momentous and rapid change.

Manyu Wana title. Pintubi Anmatjere Warlpiri Media.

Through the 1990s the activity of Warlpiri media expanded, as staff produced videos and recorded music and oral histories. They broadcast on radio and produced their own news programs that aired on local television. This diverse media work was a vital element of the larger project of 'community making' underway in this period and helped cultivate an expanding Warlpiri public sphere of discussion and debate. Some of the work produced in this period was taken up and broadcast across the region by the Central Australian Aboriginal Media Association and Imparja Television. Collaboration with professional filmmakers enabled the production of *Manyu Wana*, meaning 'just for fun', an award-winning Warlpiri-style *Sesame Street* that combined wry humour and bush-based creativity with numeracy and literacy education. In many ways *Manyu Wana* was the forerunner to *Bush Mechanics*. It was also in the making of this series that the creative partnership between Francis Jupurrurla Kelly and David Batty was formed.

Over time Warlpiri Media staff were increasingly servicing a wider region of communities with video and radio training programs and other kinds of media support. With the arrival at Yuendumu of a skilled and enthusiastic radio coordinator in 2001 the establishment of a digitally linked regional radio network became possible. The Pintubi Anmatjere Warlpiri Radio Network was set up to operate in eleven Aboriginal towns across the Central and Western deserts whose residents share close kinship ties and ceremonial relationships. The launch of the network saw an explosion of interest in radio work among budding broadcasters. Previously the broadcast footprint for these towns was restricted to within a one-kilometre radius around the local satellite dish; now DJs were able to address kin and wider audiences spread across half a million square kilometres of desert Australia, an area ten times the size of Switzerland.

The extraordinary energy around the radio network consolidated the regional focus of the media association and it soon underwent a name change, becoming PAW Media. Today PAW supports a vibrant array of media production activity – including daily radio broadcasts, video training and production, music recording, and an animation studio. PAW staff continue to record activities and events of local significance, including annual sports weekend competitions, battles of the bands, and all manner of community meetings. They also continue to make productions aimed at mainstream television audiences, at times in partnership with professional filmmakers. In recent years, such productions have included *Coniston*, which powerfully re-enacts events surrounding the 1928 Coniston Massacre from a contemporary Warlpiri perspective; *Aboriginal Rules*, a celebration of the deep Warlpiri passion for Australian Rules Football; and an animated series of *Bush Mechanics*. All of these productions are marked by inclusion of a distinctive and recognisable Warlpiri sense of humour: ironic and subtle responses to the larger settler colonial reality and politics of representation with which Aboriginal people must contend. Each of these productions provides a compelling glimpse of the way Warlpiri see their place, and their places, in a wider world of changing relationships.

HOME AND AWAY

Across the six decades since settlement, each generation of 'new media' – from film nights and two-way radio communications, to video, radio and television broadcasting, to digitisation, the internet, social media and mobile phones – has served contradictory purposes. Over a long history, recording, broadcast and digital media have been used to objectify Warlpiri along with other Aboriginal people in particular ways, depicting them as noble savages or impoverished unruly outcasts. The same media have been taken up by Warlpiri themselves to foster community and to shape a unique form of *yapa* media, and to project these self-made images to a wider world. The same media have been vital in sustaining relations between people at home in the desert and kin on the move, in hospital, in prison, travelling or residing elsewhere. Media have continuously re-oriented *yapa* to a changing world.

As I write this essay in mid-2017 a new impetus to move is very much apparent. Yuendumu and other remote Aboriginal places are in the process of being transformed in policy and public discourse from 'communities' to 'towns'. Along with this shift governments imagine they can loosen up Aboriginal people's attachments to their places, and encourage them to move to areas where it is suggested better job prospects and life opportunities are on offer. At a broader level the new impetus to move is encouraged by the relative ease of travel – physically and imaginatively. The experience of getting across the desert and into Alice Springs has been radically transformed since the recent tarring of the Tanami Highway; the once pot-holed and deeply corrugated bone-jarring ride, infamous for wrecking cars and causing dangerous accidents, now affords a smooth journey across double-lane bitumen. Rumours suggest that, in tandem with the proposed sealing of the highway from the centre to the Western Australian coast, there will soon be wireless communications rolled out right across the desert. Such infrastructure programs, whether they are realised or not, are part of a larger government vision to 'develop the north'. The permit system that once enabled Warlpiri to have some control over who entered their land and for what purposes has been removed by government legislation. Remote Aboriginal towns and Aboriginal land are envisaged as areas that can be opened up to commercial enterprise, the free market, bringing imagined new opportunities to the residents of those places.

The stimulus to move is felt on the ground in Warlpiri communities, with families now commonly dispersed across central Australia and further afield for a variety of reasons, including children enrolled in interstate boarding schools, people with chronic illness requiring intensive medical support in Alice Springs and Adelaide, and kin incarcerated in Darwin and Alice Springs prisons. Much less common are cases of people leaving in the hope of pursuing a new life. When they do, migration rarely brings with it the success imagined by governments. Isolation, homesickness and impoverishment are common.

In a time of widespread instability and social turbulence, Aboriginal people's relationships to their places are under severe pressure. In these circumstances communications media continue to work in contradictory ways – encouraging people to be open to the world, while also enabling

Above: Filming *Coniston*. Pintubi Anmatjere Warlpiri Media. Below: *Aboriginal Rules*. Pintubi Anmatjere Warlpiri Media.

Recently sealed Tanami Highway, October 2016. Photograph: Melinda Hinkson.

those who leave Warlpiri country to maintain relationships with those who stay. For those who leave, mobile phones and social media are vital carriers of news, images and music from home.

In the final episode of *Bush Mechanics* we watched on as the Jupurrurla brothers thundered across the desert and on to Broome, authorised and protected by the ancestral power of *Ngapa Jukurrpa*, Rain Dreaming, painted across their car. Travelling through the world in the company of kin and *yapa* images is an incomparable way to travel. To travel this way sustains Warlpiri ways of relating and ways of seeing – in short: ways of being *yapa*.

Melinda Hinkson is an Associate Professor of Anthropology and Australian Research Council Future Fellow based at the Alfred Deakin Institute for Citizenship and Globalisation, Deakin University.

NOTES

1 M.C. Hartwig, 'The Coniston Killings', unpublished Honours thesis, Department of History, University of Adelaide, Adelaide, 1960, p. 11.

2 A. McGrath, *Born in the Cattle*, Sydney: Allen and Unwin, 1987, p. 15.

3 M. Meggitt, *Desert People: A study of Walbiri Aborigines of Central Australia*, Chicago: University of Chicago Press, 1962, p. 21.

4 As told to Petronella Vaarzon-Morel. See P. Vaarzon-Morel (ed.), *Warlpiri Women's Voices: Our lives our histories*, Alice Springs: IAD Press, 1995, p. 45.

5 C.H. Noblett, Chief Protector of Aborigines to J.C. Cawood, the Government Resident, Alice Springs, 1 December 1928, NAA A431 1950/2768, Part 1.

6 M.C. Hartwig, 'The Coniston Killings', unpublished Honours thesis, Department of History, University of Adelaide, Adelaide, 1960; L. Watts, and S.J. Fisher, 'Pikilyi: Water Rights – Human Rights', unpublished thesis submitted for the degree of Master of Aboriginal and Torres strait Islander Studies, Faculty of Aboriginal and Torres Strait Islander Studies, Northern Territory University, Darwin, 2000.

7 As cited in L. Watts and S.J. Fisher, 'Pikilyi: Water Rights – Human Rights', unpublished thesis submitted for the degree of Master of Aboriginal and Torres Strait Islander Studies, Faculty of Aboriginal and Torres Strait Islander Studies, Northern Territory University, Darwin, 2000, p. 206.

8 D. Elias, 'Golden Dreams', unpublished PhD thesis, School of Archaeology and Anthropology, Australian National University, 2001, pp.80–83; J. Marcus, *The Indomitable Miss Pink*, Sydney: UNSW Press, 2001.

9 Yuendumu Artists, *Kuruwarri – Yuendumu Doors*, Australian Institute of Aboriginal Studies, 1986, Canberra.

10 Commonwealth of Australia, *Return to Country: The Aboriginal homelands movement in Australia*, Canberra: Australian Government Publishing Service, 1987.

11 E. Michaels, *The Aboriginal Invention of Television*, Canberra: Australian Institute of Aboriginal Studies, 1987; *Bad Aboriginal Art: Tradition, media and new technological horizons*, St Leonards: Allen and Unwin, 1994.

12 E. Michaels and F. Kelly, 'The social organisation of an Aboriginal video workplace', *Australian Aboriginal Studies*, 1, 1984, pp. 26–34.

13 E. Michaels, *For A Cultural Future: Francis Jupurrurla makes TV at Yuendumu*, Melbourne: Art and Text, 1989, p. 13.

14 E. Michaels, *For A Cultural Future*, 1989, p. 73.

15 Warlpiri Media Association, 'We been talking and talking about TV', local video recording, edit of Remote Telecommunications Meeting, PAW Media Archives, VHS video tape no. 0495, Yuendumu, Northern Territory, 1984.

THE MAKING OF BUSH MECHANICS

David Batty and Francis Jupurrurla Kelly

HOW DID THE ORIGINAL BUSH MECHANICS COME ABOUT?

David Batty: I had lived in Alice Springs for about thirteen years and I'd had a long association with Francis Kelly at Yuendumu and with the origins of Warlpiri Media Association. I decided to move to Broome and it wasn't long after that I got a phone call from Warlpiri Media in Yuendumu asking if I'd be interested in coming down to make a film about young men and cars. Now, who actually had the idea in the first place, I'm not quite sure. So I came down from Broome and had a look at it and originally they had the idea to make one short film and call it *The Mulga Clutch*. That was in response to the fabled mulga clutch incident, where Eric Michaels, an American anthropologist, came to Yuendumu and went with Francis out past Kintore and broke down on the way back. Francis made a clutch plate out of mulga, and saved the day and got all the way back to Yuendumu. So the original idea was to call it *The Mulga Clutch*, but when I got down there I had a look at it I said, 'Look, it's more about a general kind of mechanics. Why don't we call it *Bush Mechanics*?'. So I came up with the idea to call it *Bush Mechanics*, but the idea to make a film about young men and cars came from the community.

Francis Jupurrurla Kelly: We put our ideas, Jupurrurla and I talk, there's a lot of things we can do … why can't we make a bush mechanic? Because there's no garage out in communities or out on the road, we can survive to fix our own motor car, from our own skill, that's how we

Co-directors David Batty and Francis Jupurrurla Kelly. Pintubi Anmatjere Warlpiri Media.

came up. We talked with other media groups, and [decided] this is the one, we want to make a bush mechanic film to show the world how we, people, survive. We made a little clip about bush mechanics ... and they said, 'Oh, that's good!'. A lot of people, still today, they're still raging about *Bush Mechanics*, and it's good, and popular.

HOW LONG HAVE YOU AND FRANCIS JUPURRURLA KELLY BEEN WORKING TOGETHER?

David Batty: I've known Francis since the early 80s, when he first came to CAAMA [Central Australian Aboriginal Media Association], in Alice Springs, where I was working setting up the CAAMA TV unit. And about that time Eric Michaels, the anthropologist, came to Central Australia to look at the impact that the satellite would have on Aboriginal people in remote areas. He came and he met myself and my brother and Freda at CAAMA, and we introduced him to Central Australia, and told him a good place to go would be Kintore. So he went out to Kintore, he did some filming for me out there for a film I made called *Settle Down Country*, and on his return he came back through Yuendumu.

So he got to know Yuendumu quite well and he decided to make it a bit of a case study: what would happen if you set up a little TV station there. So he set up this TV station there, and before long he recruited Francis Kelly. Francis had been shooting some Super 8 with some people in Yuendumu, and he'd been around filming and photography at Yirara College when he was a student. Pretty early in the piece Francis came into CAAMA to do some video training, so he came into Alice Springs and we put him on a camera, and we did a few live OBs [outside broadcasts] and things like that, and before long we'd established a good relationship.

And then they asked me to go out to Yuendumu to run some courses for the BRACS, the Broadcasting for Remote Area Service. So I did some little courses for BRACS and showed people how to use cameras and make videos, and Francis was among those. Then I got to know Francis quite well. He came into CAAMA every so often to do little bits of training, and we gave him and Yuendumu support, and then I went out and made *Manyu Wana*, in the late 80s. I did it over four years: '88, '89, '90 and '91. During that time Francis was involved in it, and he was an actor in some of the scenes we did in *Manyu Wana*. So we go back to the mid 80s and we've collaborated on a lot of projects since that time.

Francis Jupurrurla Kelly: Well, in 1985 we went across to Kintore, and talk about medias like people bringing in satellites and all that, and how were we going to cope with that because we want to teach our young people through that. Now, today, I can see every community got their own media, to cope with it and to operate their systems how they want it. It was Freda Glynn and Philip Batty they're the two that were working for CAAMA and teach us, and David Batty, Jupurrurla, he was with us out in the field, getting all the stories together.

We know each other pretty well because he knows the skill, I know the skill. We've been sharing it, how to do it, how to make proper films about Aboriginal people, and about the language, because he's got good sense of humour, too. In our society, David Batty is Jupurrurla. Same skin name: Jupurrurla to Jupurrurla. We are brothers. We respect each other. Whatever he do, I help him. Whatever I do, he helps me. That's what it means. Skin groups are family.

David Batty films Francis Jupurrurla Kelly. Pintubi Anmatjere Warlpiri Media.

HOW DID *BUSH MECHANICS* BECOME A SERIES?

David Batty: Well, it was the very next day after the first *Bush Mechanics* went to air. I got a phone call from Stefan Moore who was at Film Australia at the time. He was an executive producer. And Stefan was a New Yorker and he contacted me in Broome and his words were something like 'David Batty, I just saw *Bush Mechanics* on TV and I fuckin' love that show, man!' And then he said, 'Can you do four more?'. And I thought for a little while and I thought, yep, we can do four more, and I rang up Francis and got hold of him and said 'Francis, do you want to do four more episodes?'. He was keen, so a little later I drove down to Yuendumu and got together with the bush mechanics from the very first show and we had a little meeting and I put it to them. I said 'Now, are you guys keen to do four more episodes of *Bush Mechanics*?' and they were extremely keen. So that's how the next series came about.

WHAT WAS IT LIKE TO TRAVEL INTERNATIONALLY AFTER THE SHOW?

David Batty: Francis wasn't keen on travelling overseas so I just represented it myself. We had a screening at Hot Docs in Toronto in Canada soon after the original *Bush Mechanics* was made. We showed *Bush Mechanics* in a little cinema in downtown Toronto and we had a packed house and there were quite a few people who had worked with the Inuit people and some Inuit filmmakers and they really got it. They absolutely loved it. They laughed at all the right spots and the audience response was really phenomenal. That was when I realised that we had a bit of an international hit on our hands. It has since travelled internationally, and it was very gratifying to have it seen by so many people all over the world.

WHAT DID OTHER PEOPLE IN YUENDUMU THINK ABOUT YOU MAKING THE SERIES, AND WHAT ABOUT THE MEDIA ORGANISATIONS YOU APPROACHED, LIKE THE ABC?

David Batty: When I went down to make *Bush Mechanics*, people knew me quite well because I'd made the series *Manyu Wana* ten years prior. *Manyu Wana* was such a success; the children had really loved it, and it had become a real house movie that the kids just loved to watch over and over, because it was all in Warlpiri. *Manyu Wana* was created as a response to the kids potentially losing language and being bombarded by mass media from the satellite, *Sesame Street* and all those things. Over four years I made ten little half-hour films, so people knew me; they knew *Manyu Wana*. I was very much involved with the community. I had been given a skin name; people had no problem with me doing any filming out there. I'd been involved with other projects out there as well.

Francis Jupurrurla Kelly: With *Manyu Wana* I was talking to David Batty about, you know, teach young people in language about counting numbers, in the school. Me and him; he came from CAAMA and I was at Warlpiri Media. We both worked together and made little clips about counting numbers, like, *jinta, jirrama, marnkurrpa*. That's in language; it's one, two, three, four, and goes up to ten, in language. And from there it's about not only counting numbers, but learning kids about ground paintings and drawings on the ground, where old people tell them, you know, instead of reading books. In those days, there was no books, but they used to draw on the ground and tell the young people, young children, to learn about cultural sites.

David Batty: So when we came to make *Bush Mechanics* a lot of people saw it as an extension of me coming to make *Manyu Wana*. That's one of the reasons I employed the same techniques, using the wind-up Bolex camera, and shooting in mute and then editing it in post-production and getting the *Bush Mechanics* back to voice the sections that were edited to give that kind of effect. It's just the way it evolved, really. I didn't think about it too hard. It was the way *Manyu Wana* evolved: out of necessity, through lack of money to do anything else! So I employed the same techniques in *Bush Mechanics*, and people were familiar with it. They were familiar with the Bolex; they were quite fond of the Bolex. They knew what it meant when I said they had one go at something. It was a twenty-second shot and that was all you could get. You couldn't sit there and film all day. People knew that I knew all the various cultural protocols – things to do and what not to do – and I knew a lot of the elders at that time. I'd worked with them in the film at Warlukurlangu Artists, so it was an easy fit for me. I think the community were very happy.

Brian McKenzie at the ABC was a staunch advocate for *Bush Mechanics*. He's a documentary filmmaker and was commissioning editor at the ABC – from the very first one we made he absolutely loved *Bush Mechanics*. He was a very strong supporter of coming back to do the series. We just had doors opening everywhere we went with *Bush Mechanics*, to come and do the series. And it got prime-time spots on ABC. It was really well supported – lots of publicity. It just seemed to really hit a nerve with the viewers, and it was just what people wanted at that time. The ABC loved it.

Right: Shooting the *Bush Mechanics* series. Pintubi Anmatjere Warlpiri Media.

Below: The bush mechanics. Pintubi Anmatjere Warlpiri Media.

YOU AND FRANCIS ARE BOTH BUSH MECHANICS BUT LEARNED IN DIFFERENT WAYS. WHO TAUGHT YOU AND HOW?

David Batty: I grew up down in Wollongong on the south coast of New South Wales. My father was a very keen bushwalker and canoeist and we spent a hell of a lot of time in the bush. He was also very handy. He could fix most things. My grandfather was an inventor and a time-and-motion study man and could make and fix and invent. He had a really different take on things. He was incredibly ingenious. I always loved pulling things apart, and fixing them, and putting them back together. And from quite a young age I never really had much money and so I had to work on my own cars. And then moving to Alice Springs it was more of the same. Building a car, and fixing utes, and changing motors and getting stuck long-distance travelling, just trying to work out a way to get the car going to continue on your journey.

Also, for me, when I was a teenager we were very heavily interested in old British motorbikes so we would get these old motorbikes, pick them up for a song. Lots of them were ex-army motorbikes and we'd strip them all down and put a straight-through pipe on them. We'd take off all the guards. We'd take the seat off. We'd put on a little lawnmower petrol tank. We'd strip off all the chain guards. They were all pre-unit which meant that you just had the donk with the chain going to the gear box, and the clutch and all the clutch plates – they were all exposed, and they'd be chugging away and they were as dangerous as all hell, and we'd make a little seat out of a piece of wood, put a bit of foam on it. And we'd put on big white handlebars, and if some of the boys had a bit of money they'd save up and get a knobbly tyre, and put that on the back. And these things were like 500 singles, 650s, 500 twins, Triumph speed twin. We had Beezer 500s, Ariel 350s – all these old British bikes we used to take up behind Wollongong in the bush and we'd just tear around on those. We'd just tear around on those all weekend and we'd just camp up there and these bikes were our life. We just loved them so much. That's really where I learned a lot of my mechanics, on those old motorbikes.

Then when we got to cars it was just an extension of that, and we actually did the same with cars. We used to pick up cars on a Friday afternoon from people's backyards. We'd get them for $5 and $10 and they were all old rust buckets, cars that today are worth a fortune: Morris Minors, Woody Wagons, all the early Holdens. We had Oxfords, Hillman Hunters. So we'd do the same with those. We'd manage to get them going. There would always be something wrong with them – that's why they were sitting in people's front yards on blocks for so long – but we'd get those old cars and we'd take them up Mount Keira and over to Mount Kembla and we'd just tear around and they'd only last a weekend and by the time we'd finished with them they were completely trashed. So my youth was spent with motorbikes and old cars around Wollongong. And that's really where I learned a lot of my bush mechanics. Then, as I grew older, it was more out of necessity and trying to find ingenious solutions to things when you get stuck in the bush.

WHY THE JUPURRURLAS? WHY ARE THEY THE STARS OF THE SHOW?

David Batty: It was just the way things panned out. When I came down to make *Bush Mechanics* I was there for a week or two and we'd been talking about what we were going to do, particularly with Simba, who's a Jupurrurla, and Francis, who's a Jupurrurla, and I'm a Jupurrurla, which means we're brothers. I'd got hold of an old car, and I'd been mucking about with that for a while, trying to get it going. People didn't really know what I was up to, and I didn't really know what was going to happen and how it was going to pan out. We had a meeting at Warlpiri Media, and we put it out there, calling on young fellas who wanted to be in a TV show, and it was a Friday afternoon, so about ten young men turned up, keen as anything. We told them the story. We had decided that the story would be that the guys, four young men, drive the clapped out car to Alice Springs to buy a new car, and drive home. That was the story. And then we would sprinkle through it some of the bush tricks, in the *Manyu Wana* style.

Friday afternoon came along, these ten young fellas rocked up, we had a meeting, they all said they were keen, so I said okay, the first four young guys who turn up here on Monday morning, they can be the *Bush Mechanics*. We're definitely going Monday morning and we'll be filming this TV show. Monday morning came and not one of the young fellas turned up. So I said to Simba [Adrian] Nelson, who had been working at Warlpiri Media and who had been part of this project, 'Get in the car and go around and get three more. You're going to have to be one of the bush mechanics instead of being a co-director or having a production role'. So Simba got in the car and drove around the community. We're all waiting there all set to go – we had tucker for the road, we had the camera, everything we needed – we just needed three more young guys. Simba wasn't gone that long, about an hour, and he came back with these three guys all scratching their heads, 'What is it, Jupurrurla?'. And of course because it's an obligatory relationship he just went straight to other Jupurrurlas, to his brothers. He knew he could ask them and they would have to comply. So that was how they came to be all Jupurrurlas, and that's how they came to be bush mechanics.

HOW'S LIFE CHANGED AFTER THE SERIES? WHAT HAS SINCE HAPPENED TO THE JUPURRURLA GANG?

David Batty: Well, life for me personally changed because of the success of *Bush Mechanics*. My filmmaking career took off after that. My collaboration with Francis blossomed and we went on to make more shows. I think for Francis, he definitely become a well-known character, not only within his community, but throughout Central Australia and then throughout Aboriginal Australia. Francis is incredibly well known. When you travel with Francis throughout Central Australia everyone knows him. I think it really helped put Warlpiri Media on the map. *Bush Mechanics* put them in a much stronger position to apply for money, to employ more people, to become quite a strong voice in the community and to continue their work creating media, with their radio and also their TV.

The bush mechanics. Pintubi Anmatjere Warlpiri Media.

And what's happened to the Jupurrurla gang? Well, one of them seems to spend a fair bit of time in and out of jail. Stephen's moved to Lajamanu, and is living up there with this family – he seems to have a pretty good job up there. Junior continued his football career and seemed to travel around to all the different communities and keep playing football. I don't know what other career he pursued. Simba continued working at Warlpiri Media. He was the mainstay of Warlpiri Media from the very first *Bush Mechanics* up until he passed away just a few months ago. Francis, of course, went on to bigger and better and brighter things. We collaborated to make *Coniston*, the one-hour documentary about the Coniston Massacre. Francis was involved in various other film projects and things but, of course, now he's gone on to become Chair of the Central Land Council which is an incredibly prestigious and incredibly demanding job. I never would have thought that a little show out in the bush would have such a far-reaching impact on so many people's lives and on the community of Yuendumu. It never ceases to amaze me.

ARE THE TRICKS REAL? WHAT'S YOUR FAVOURITE ONE?

David Batty: Are the tricks real? Of course they're real. All the tricks in *Bush Mechanics* are real. Some are more real than others . . . they're all derived from either things that Francis has witnessed or had to come up with on the spur of the moment to keep a car going and get himself out of a pickle or things that I've seen on the road. In earlier days if you spent enough time travelling around the bush everywhere you went you'd see broken down cars and you'd see the most amazing fixes on the cars that people had come up with using bits of mulga and ubiquitous wire. I can't think of any of them that are too far-fetched to be real.

I think one of my all-time favourites was one that Francis came up with. When your electrics are playing up and your coil is getting too hot – well, then you get a billy can or empty fruit tin, and you make that into a little billy. You put a little bit of wire through it to make a handle, and then you put the coil into the tin can, you put spinifex around the outside, and then you fill it up with water. The spinifex is to stop it sloshing around too much and so that's like a water cooled coil which cools the coil and stops your electrics from overheating. That's the one I love the most. Another one I particularly like, which is in the show, is the windscreen wipers, when the wipers have gone and you wrap bits of cloth around the wipers to do the same job. But they're all real and they're all little tricks to get you out of a pickle.

WHAT WAS FUN, AND NOT SO FUN, ABOUT SHOOTING?

David Batty: For me, the fun times were when everything stopped and we camped out with the boys in the evenings around the fire. They were some of the most hilarious moments in my life, when Francis, being the true comedian that he is, would keep us all enthralled with stories and re-enactments of things that had happened to him. He's the most hilarious storyteller. I can remember times when we would all be sitting around the fire literally rolling around in the dust, crying with laughter. The other thing would be endless jokes about skin names and making fun of other skin names. Being all Jupurrurlas the conversation would often swing around to putting down the other skin names, and Japaljarris for some reason seemed to really cop it. The other fun times were when we'd come back from shooting, we'd have a barbecue going, and we'd have a fridge and freezer full of fresh meat from the nearby station. And we'd cook up and have these huge feeds from the barbie, and then we'd watch the rushes from that day. Various family members, wives and kids, would all crowd into this tiny little place we had – all crowd into this room and watch the rushes. Simeon was a comedian as well, and just as soon as he came on the screen there would be this uproar of laughter and everyone crying out for Simeon – they called him Mr Bean. He really was like the Warlpiri Mr Bean; as soon as he came on screen you just had to laugh. Whatever he did, it was just funny. So I think it was the times that we weren't filming that were the funniest.

The time that was the toughest for me was when did the final episode, 'The Rainmakers', and the trip to Broome where the gearbox on the Ford we had blew up. It was pretty early in the piece, the second day of quite a long journey to Broome, the gearbox had gone, and it really pushed me to my absolute limit, in terms of trying to work out how we could keep filming with a car where the gearbox didn't engage with the motor. It meant we had to get shots where we'd shunt the car with the bull-bar of the Toyota. This was right out in the middle of the Tanami Desert, no support, no one around, no one to help us, and we just had a small crew. We had the boys in the Ford, Francis in the Toyota and me in a Toyota. So we had to do things like shunt from behind with the bull-bar of the Toyota, push the Ford, get it up to 100kms an hour, and I'd be off filming in the distance getting a long shot, then they'd slam on the brakes and the Ford would just be rolling, rolling along the desert. All the travelling shots of the Ford coming up the Tanami in 'The Rainmakers', the car's not actually running. We ended up shunting it, pushing it, towing it, and the last few hundred k's of the Tanami Track – which is a pretty rough road, corrugated, dust, rocks, dirt, very remote – we just towed it with a tow rope. And when you tow a vehicle on a tow rope on a bad road like that all the rocks get thrown up from the car that's towing the vehicle, so consequently the windscreen got smashed, all the lights got smashed, everything on the front of the Ford got smashed by rocks flying up from the car that was towing it. That created more difficulties, because we had to get from Halls Creek to Broome without a windscreen, so we found one in Fitzroy Crossing, which was slightly staged. We had to arrange to get a gearbox in Halls Creek. At that time all I had was the little Flying Doctor radio, a green tin box where you could ring anyone in the country via the Flying Doctor. You had a three-minute call. It was a queue system, like a party line, so you'd get up in the morning and the Flying Doctor would come on, and you'd get out your little green tin, tune it in, throw a wire over the nearest tree, lift your bonnet, connect the wires up to your battery, put the clips of the wires onto your radio, and tune in and book a call through the Flying Doctor. So it was via this radio that I managed to get a gearbox sent from Alice Springs out to Yuendumu, and then I got a plane to fly down from Halls Creek, take the seats out of the back of the little Cessna, fly down to Yuendumu from Halls Creek, load the gearbox into the Cessna, and fly it back up to Halls Creek. So by the time we actually got to Halls Creek we had a second-hand gearbox to put into the Ford. Unfortunately the boys decided to go and party on and drive around Halls Creek, while me and Hugh Miller, the sound recordist, and Davey, one of the drivers who became one of the bush mechanics, we worked all night, right through till dawn, taking the broken gearbox out of the Ford and putting in the one that was flown up from Yuendumu. That wasn't much fun. That was probably the part that I found the most difficult, through the whole making of the series.

ARE *BUSH MECHANICS* TRICKS STILL PART OF EVERYDAY LIFE?

David Batty: Bush mechanics are still a part of everyday life but to a much lesser extent, I think mainly because cars are harder to work on these days. They're a lot more electronic. They don't have carburettors. They have electronic ignition. Things that were quite simple mechanical ideas and devices and concepts under the bonnet are now much more complicated and you need computers to fix things. It even comes down to, some of the new cars now once you've lost the key you can't trigger the electronics to start the motor. Recently I've been to communities where there're rows of quite new cars where people have lost the keys and the cars are completely useless. They can't do anything with them. That's something they can't fix. The other thing that's happened is there're a lot more bitumen roads, like the roads to Yuendumu now are almost all completely bitumen, so the cars are not falling apart as rapidly. They're not suffering from endless amounts of to-ing and fro-ing to places on corrugations, so I think the requirement for people to be skilled bush mechanics has diminished somewhat but people still know how to fix the older cars. They like older ones. They actually do like the cars like the older Fords and the older Holdens where they do have carburettors and they can fix and repair those cars but I think it's become less a part of everyday life. The other thing is that cars now are so much cheaper and people have more money. I think when we did *Bush Mechanics* cars were a scarcer commodity. You needed a fair bit of money to buy a car back in those days. But now cars are cheap, roads are better, and cars are more complicated so I think there is a diminishing in the skill base of people in the bush knowing how to fix things.

David Batty answered a series of questions put to him by Mandy Paul and Michelangelo Bolognese. Comments by Francis Jupurrurla Kelly were sourced from the ABC's 2016 Australian Story *episode about David Batty.*

NYURULYPA

The good tricks

Michelangelo Bolognese

Outlandish fixes that seemed unlikely to work were probably the element that elevated Bush Mechanics to cult status. And while they may seem an improvised series of tricks, they are not unique to the characters of the show nor even to their country. Bush mechanics have existed in Australia longer than cars have. Workers on remote stations across Australia had to operate and fix machinery by themselves, without workshops or specialised equipment. Many Aboriginal people in Yuendumu and all over Australia were capable bush mechanics. When cars arrived, these mechanics quickly adapted their skills to the new technology. Far from spare parts and sophisticated tools, they kept their cars running with what was available in the bush. Mulga wood, spinifex and sand are some of the things that make up the bush mechanic's toolkit. A bush mechanic must have a deep understanding of mechanical systems, a measure of ingenuity and the capacity to improvise. Some of the tricks (nyurulypa in Warlpiri) seen on the show are bush mechanic staples – well-known solutions to common problems. Others are unique responses from the Jupurrurlas to problems they encountered during filming or earlier in their mechanical careers. This chapter outlines nine of their best nyurulypa and explains the way that they work.

SPINIFEX TYRE

Tyres are the point of contact between cars and the road. The rotational force of the wheels is translated to forward motion on the ground at the point of contact. The first car wheels were wooden, but rubber tyres (already used on bicycles) were quickly adopted due to superior handling and suspension qualities.

A punctured tyre is the most common failure in a car. The air inside the tyre will escape the chamber, and the flat tyre results in a bumpy ride and poor handling. Cars carry spare tyres as standard accessories, and a flat can be repaired in minutes along the roadside.

If a spare tyre is not available, a perfectly sound bush mechanic's fix is to stuff the tyre with spinifex. This is a dry, spiky grass found throughout Central Australia. After removing the tyre's inner tube, tufts of spinifex densely packed inside the damaged tyre can create sufficient pressure inside the chamber to enable the car to be driven safely.

WIPER FUEL PUMP

In cars powered by petrol or diesel, the fuel pump is a small pump connected to the carburettor (in cars without electronic fuel injection). This is where fuel is mixed with air and delivered into the cylinder to be ignited in the process that is at the heart of combustion engines.

If a fuel pump has failed, the engine simply cannot run. Fuel pumps are regularly replaced during a vehicle's service schedule to minimise the likelihood of a breakdown. They are normally relatively inexpensive parts, since the pumps are simple and small, and a mechanic can easily replace a pump with a new one.

It is unlikely that a pump can be repaired on the roadside, since a failed pump will normally have sustained extensive physical damage. An ingenious bush mechanic hack is to replace the fuel pump with the pump fitted in most cars to send water to wash the windscreen. Though having a clean windscreen is an important part of road safety, a car can run without this part. If a hose is run from this pump directly to the carburettor and the water in the reservoir replaced with fuel, pushing the right button will send fuel to the engine. While the car may not run smoothly, this trick will allow the car to limp along.

BOOMERANG CLUTCH

The clutch is the component that allows gears to be changed smoothly and without damage. The constant upwards and downwards motion of the pistons in a running engine results in a rotation of the crankshaft, but the optimum rotation of the wheels relies on the right gear being selected. Gears are toothed metal discs that fit into each other, and when gears are changed they would simply crunch each other if transmission of power from the drive shaft to the gearbox was not momentarily suspended. This is the job of the clutch, and all cars with manual transmission are fitted with clutches. There are a number of different types of clutches in cars, but most rely on springs and friction plates to halt the transmission of power from the drive shaft.

Jack Jakamarra Ross and the boomerang clutch. Pintubi Anmatjere Warlpiri Media.

A clutch failure is a debilitating problem, making a car extremely difficult to run, as selecting gears without a clutch is almost impossible. The main cause of failure is excessive wear – particularly common in cars with drivers who 'ride the clutch'. Though replacing the actual clutch plates is relatively simple, removing the clutch to access them is often a laborious process and is normally carried out by skilled mechanics. A broken clutch can be an expensive mishap!

One of the most brilliant *nyurulypa* was shown in the original *Bush Mechanics* documentary, and features the 'boomerang clutch'. After the clutch failed, it was removed from the car and a new clutch plate was needed. Elder Jack Jakamarra Ross carved two little boomerang-shaped pieces of mulga that were fitted into the clutch, which worked well enough for the car to limp home. Director David Batty recalls that the scene is a recreation of a trick devised by co-director Francis Jupurrurla Kelly while out on a trip with anthropologist Eric Michaels. Francis's 'boomerang clutch' was later removed and put on display in a frame at the Australian Institute of Aboriginal and Torres Strait Islander Studies, in Canberra.

BRAKE FLUID

Though a few different types of braking systems exist, the most widely used is a hydraulic system. Hydraulic brake systems make use of the properties of pressurised brake fluids to transmit the braking command from the foot pedal to the wheels. Special fluids are required that can sustain the high temperatures generated in the brake system and have the right compressibility and viscosity.

Leaks in the brake lines or wear to the brake pads are common causes of a drop in the level of brake fluid. It is important to keep brake fluid at a good level, as a car with insufficient brake fluid will have brakes that are less responsive or may stop functioning altogether. A catastrophic failure of the brakes can be very dangerous indeed! The correct type of brake fluid for a car can normally be found easily and inexpensively, and topping up the reservoir is an easy task.

If a brake failure happens in the desert and the brake fluid needs to be topped up, it is unlikely that a bottle of specialty fluid will be hiding behind a tuft of spinifex. A *nyurulypa* commonly used by bush mechanics, however, just requires a bit of common laundry powder. Mixing a few spoonfuls of laundry powder with water will result in a viscous liquid that will return the hydraulic system to operation. If you've had to resort to this trick, it's a good idea to flush the laundry powder brake fluid and replace it with the correct fluid as soon as possible, as water will cause the brake lines to corrode quickly.

FIRE STARTER!

The battery is an essential component of a car: as almost anyone knows, a car will not start with a battery that is not charged. This is because in most cars with combustion engines the battery powers the starter motor which in turn begins the 'four-stroke' cycle. This cycle requires an external power source to kick it off, as each cycle relies on the inertia of the previous one to initiate the next. The battery may also power certain components once the car is running, like the Electronic Control Unit (ECU) in many modern cars.

A dead battery is probably the second most common of all car problems. Usually simply caused by forgetting to turn off one of the electrical components (like the headlights) after the engine is stopped, there are a few typical solutions. Engines can be 'jump-started' by connecting them using leads to a charged battery (normally one connected to another car). This allows the car with the dead battery to draw power from the charged battery and begin recharging its own battery through the alternator once the engine is running. Another is to 'push start' the car: building up enough speed by pushing the car and quickly engaging second gear can create enough momentum in the pistons to initiate the power cycle, which will again begin to recharge the dead battery. If neither of these methods is an option, batteries can be removed and recharged or replaced.

Jump-starting and push starting are often impractical in the desert, as the first relies on the presence of another car while the second requires sufficient momentum to be built up, which is hard on a sandy unsealed road. A common bush mechanic fix is to remove the battery and place it for some time on a fire. The heat from the fire transfers energy to the electrons and protons in the battery, which become excited and give the battery another short burst. This can be enough to start the car, but placing a car battery on a fire can be very dangerous and is not for the faint-hearted.

RADIATOR PATCH

A combustion engine essentially works through a steady series of controlled explosions within its cylinders that, through a sophisticated train of components, delivers power to the wheels. It is no surprise, then, that engines get very hot! Sustaining operation under this heat would be impossible without a cooling system, so cars have had radiators since the very beginning. Karl Benz, whose 1885 Motorwagen is commonly understood to be the first car ever built, invented the water-filled radiator to control the temperature in his engine. Radiators work by having a coolant circulate through their cores which have a large surface area for the surrounding air to cool the coolant through heat exchange. Though most modern cars use specialised coolants, water is a common coolant in older vehicles.

A crack or leak in a radiator can be a debilitating problem. Resulting from general wear, engine stress or impact, cracks or leaks can allow the coolant to escape and reduce the radiator's cooling capacity, which will cause the car to overheat. Radiators are normally inexpensive and can be replaced by mechanics, though cores can often be replaced or cracks repaired if they are not too severe.

In the extreme heat of the Central Desert it is not uncommon to have engines overheat. If a radiator fails in an isolated location, any roadside assistance may be very far away. Though making repairs to the radiator under these conditions may seem unlikely, there is a clever (though dangerous) bush mechanic hack. An old battery can be broken down and the lead inside melted down in a hubcap over a fire. The molten lead, poured over the crack, will harden and patch the leak. This trick is shown by the bush mechanics in the 'Payback' episode while chasing the Yuelamu mob who have stolen the Yuendumu Magpies' footy coach's car.

WIPER PADS

Driving during a downpour without windscreen wipers is uncomfortable and dangerous, so it's no surprise that modern wipers were invented shortly after the popularisation of motorcars. They were first patented by American Mary Anderson in 1903. Activated manually at first, the first electric wipers arrived in the 1920s and this excellent technology has remained largely unchanged since.

Though worn or missing wiper pads may sound like a relatively harmless problem, it can greatly impede visibility and result in some nasty situations. Luckily, wiper pads are easily replaced and normally inexpensive.

Bush Mechanics shows a solution to missing wiper pads that does not involve the 600 kilometre return journey from Yuendumu to Alice Springs. Small strips of blanket can be cut and wrapped tightly around the wiper blades, providing an absorbent and smooth surface to clear any water from the windshield.

GETTING UNDER THE CAR

When buying a second-hand vehicle, most people will look under the bonnet (many not knowing exactly what they're looking for). But while lifting the bonnet gives access to the systems most essential to regular maintenance, more involved maintenance and repairs often demand access to a vehicle's underside. This is where parts like gearboxes, differentials, axles and CV joints normally reside.

The simplest way to access the bottom of a vehicle is to use a jack. These come in various shapes and sizes, and most cars will carry a small mechanical or hydraulic jack to lift one side of the car high enough to change a wheel. Safer and more comfortable than jacks, however, are pits (below ground level, in which

The bush mechanics at work. Pintubi Anmatjere Warlpiri Media.

55

Shooting the *Bush Mechanics* series. National Film and Sound Archive of Australia 134795.

one can stand beneath a car's underside) and hoists (which lift the entire car above head height). These are found in most professional – and many amateur – workshops.

Bush mechanics often need to get underneath a car and can rarely count on the luxury of a pit or hoist. Luckily, there are several tricks to access the underside of a car. A solid jerry can wedged under the car's floor can act as a stand. Better yet, a car can be driven or pushed along a stump of mulga that is propped as a ramp on an anthill. These structures, built by certain termites, are tall enough and strong enough to support a car so that bush mechanics can slide underneath it.

COLLAPSED ROOF

Although we nowadays think of a roof as an integral part of most cars (with the exception of sporty convertibles), early automobiles were mostly roofless. As metal fabrication techniques improved in the early twentieth century, fully enclosed metal cabins became common, offering a higher level of comfort. Many cars can be fitted with roof racks and additional luggage compartments above the roof to increase storage capacity.

A collapsed roof is a severe and rare problem, and usually results from an accident in which the vehicle has rolled over. The structural integrity of the entire body may be compromised by a collapsed roof, and no amount of panelbeating will return the car to a road-legal condition.

A clever solution to a collapsed roof is shown by the bush mechanics in the first episode of the series. Overloaded with musical instruments and amplifiers, the roof caves in on the bumpy road between Yuendumu and Willowra. Armed with an axe, the bush mechanics hack off the roof and attach it to the rear of the car with a rope – a makeshift trailer. Definitely not road-legal, this trick is only likely to work on a sandy dirt road, as the trailer would not slide comfortably on a bitumen surface.

BUSH MECHANICS

The exhibition

Michelangelo Bolognese and Mandy Paul

We left our old camp, and were hunting as we walked toward Wayirdi. It was there that we came across the strange track.

The original *Bush Mechanics*, a half-hour documentary that first screened in 1999, immediately gripped viewers with an amusing recreation of how the Warlpiri people of Australia's Tanami Desert first encountered a motorcar. The 'strange track' that Jack Jakamarra Ross spots as a young man (his young self played by Francis Jupurrurla Kelly) while hunting for kangaroo is the tread of a truck. When the action switches to the present day, young Warlpiri bush mechanics are confidently driving, breaking and fixing their own cars. The documentary was followed in 2001 by a series of four episodes of the same length made in Yuendumu by writer/director David Batty and co-director Francis Jupurrurla Kelly, and produced by Warlpiri Media Association in association with Film Australia and the Australian Broadcasting Corporation. When it screened on the ABC in 2001, the series was a national hit, watched and loved by over three million Australians. *Bush Mechanics* was something new: spoken mainly in Warlpiri and subtitled in English, the series combined humour, engaging characters and everyday life in the bush. The stories centred around cars but provided much broader insights into contemporary Aboriginal life.

The special something of *Bush Mechanics* – its ability to make people smile, the power of its insights presented from a Warlpiri point of view – has not diminished over time. Nor has the need for cross-cultural understanding and connection. Introducing *Bush Mechanics* to a new generation of audiences through the medium of a museum exhibition was newly employed

Francis Jupurrurla Kelly in character. Pintubi Anmatjere Warlpiri Media.

National Motor Museum curator Michelangelo Bolognese's idea. Mandy Paul, then a senior curator at the History Trust of South Australia, had lived and worked in Central Australia, and had enough local knowledge, and experience in the museum sector, to be confident that it could work. We thought that *Bush Mechanics* had potential to be a popular, engaging exhibition, one that was centred around cars but delivered much more.

The National Motor Museum in the Adelaide Hills holds Australia's most important collection of motor vehicles and material culture related to motor transport. The museum is also a research centre, a place where stories of the vehicles in its collection, and, more widely, of Australia's motoring history, are collected and shared.

The museum has taken cars on the road before. In 2008, a treasure from the museum's collection, a 1908 Talbot tourer, was the centrepiece of the exhibition *Off the Beaten Track: A journey across the nation*. This exhibition commemorated the centenary of the first continental crossing by a motor vehicle. The Talbot travelled in a special exhibition trailer, following as closely as possible the original journey from Adelaide to Darwin undertaken by Harry Dutton and Murray Aunger, and visiting twenty-three towns and twenty-one schools along the way. The tour succeeded not only in acquainting people with the history of Dutton and Aunger's journey, but also in providing a contemporary perspective on that journey.

The warm welcome the Talbot tour received from communities across South Australia and the Northern Territory demonstrated the widespread interest in stories about motoring history and culture. And the tour itself established a new model of taking significant vehicles to locations outside the country's major cities.

The Australian outback holds a particular place in the non-Aboriginal imagination, one that the Talbot's crossing of the continent in 1908 helped to construct. In the context of motoring history, the outback is usually portrayed as a hostile environment to be conquered, and in this narrative Aboriginal people have only walk-on parts. The National Motor Museum took the opportunity offered by the 2008 Talbot tour to reflect on the relationship between cars, colonisation and the bush, and to engage with Aboriginal communities along the route.

The development of *Bush Mechanics: The exhibition* offered an opportunity to work closely with an Aboriginal community and an Aboriginal organisation around a motoring theme, and in a context where Aboriginal people played a central role. An exhibition which takes as its starting point Yuendumu's *Bush Mechanics* is underpinned by the characters' desert knowledge and familiarity with their country, inverting the non-Indigenous 'outback' paradigm. A travelling exhibition would also make possible the return of two cars from the series, now held in the collections of the National Museum of Australia and Museums Victoria, to their home communities.

DEVELOPING THE EXHIBITION

With the help of anthropologist Melinda Hinkson, we contacted Pintubi Anmatjere Warlpiri (PAW) Media to float the idea of a travelling exhibition based on *Bush Mechanics*. It was warmly received by staff, so a proposal was drafted to go to a meeting of the PAW Media board of directors. The proposal was approved.

The success of an application to Visions of Australia for development funding enabled the initial scoping of the exhibition. Research into *Bush Mechanics*, Aboriginal media production and the broader history of Aboriginal people and cars, including the work of Melinda Hinkson and Georgine Clarsen, followed. A research trip to Yuendumu to visit PAW Media headquarters and have a look at the archives was productive, and there we met Jason Japaljarri Woods, who agreed to create art design elements for the exhibition. Meeting Jeff Bruer and PAW Media's Chief Executive Officer Michael Taylor in person after many phone calls helped cement the co-operation between the two organisations.

Another key contact was David Batty, who was working on *Black As*, a show that features Aboriginal and non-Aboriginal men working on cars in Arnhem Land. Initial discussions with the National Museum of Australia and Museums Victoria indicated that each institution would consider lending the *Bush Mechanics* car in their collection, which meant that a small but far-ranging travelling exhibition could be built around these feature exhibits.

During the development phase the exhibition took on a life of its own, and enthusiasm for the project was infectious. Hours on the phone to places across Central Australia and the Top End yielded a list of venues keen to host the exhibition. A well-developed application for touring funds was successful, again through the Visions of Australia program.

The final realisation of the exhibition was completed between January and April 2017. Over a year of research, conceptual work and logistical planning led to a small conference room in Alice Springs on a sweltering February morning. A brief silence followed the end of the presentation on the proposed exhibition concept to the board of PAW Media. The pause was nerve-racking: the entire board was present, and there was no back-up plan if they did not give approval for the exhibition to go ahead. 'Youngfella,' a board member finally said, 'this is very good: you go and make this exhibition'. Soon all of the board members, who are mostly prominent senior men and women from Yuendumu, were offering up advice on details of the objects, content and even the graphic design of the exhibition. They were enthusiastic about the prospect of the exhibition's tour – not only because the fondly remembered cars would finally make another appearance in Yuendumu, but also because of the hundreds of thousands of visitors expected to engage with *Bush Mechanics*, now almost twenty years since it was first shot.

Exhibition curator Michelangelo Bolognese arrived back in Adelaide with the go-ahead, an approved exhibition text, an esky carefully packed with the *Bush Mechanics* claymation characters, and a container of red desert sand.

The exhibition opened at the National Motor Museum on 11 April 2017, and the event was attended by a strong contingent from Yuendumu, including one of the original *Bush Mechanics*, Randall Jupurrurla Wilson, cultural advisor Simon Japangardi Fisher, David Batty, and Jason Japaljarri Woods and Jeff Bruer from Pintubi Anmatjere Warlpiri Media.

BUSH MECHANICS: THE EXHIBITION

From the start of its development, one of the exhibition's main objectives was to introduce the original *Bush Mechanics* show to a new generation of Australian and international museum visitors. For this reason, the exhibition is rich in original footage, beginning at the entrance, with a specially commissioned extended trailer of *Bush Mechanics*. Large projections and smaller touchscreens feature more original scenes, while a large digital touch table showcases some of

Bush Mechanics: The exhibition. Photographer: Andre Castellucci. National Motor Museum.

the best *Bush Mechanics nyurulypa*, or tricks, (as selected by David Batty, and accompanied by his commentary). These elements accompany more hands-on interactives, including a 'bush driving simulator' developed by self-described 'tinkerer' Mark Thompson that can only be played after a steering wheel is found among bits and pieces in a nearby toolbox. There is also a 'claymation studio' (developed in partnership with Adelaide-based studio GooRoo Animations) where visitors can create their own plasticine stop-motion animation. These interactives encourage visitors – children in particular – to engage with the exhibition and to understand its content through doing and making, just as bush mechanics' repairs are learned. Importantly for its development team, which was made up of a number of people based mainly in Adelaide and Yuendumu, the exhibition is aimed at both Aboriginal and non-Aboriginal visitors, and the exhibition text is presented in English and in Warlpiri languages.

Above: The driving simulator. *Bush Mechanics: The exhibition*. Photographer: Andre Castellucci. National Motor Museum.

Below: A digital touch table. *Bush Mechanics: The exhibition*. Photographer: Andre Castellucci. National Motor Museum.

The exhibition is designed to capture the energetic and upbeat tone of the show. PAW Media artist and claymation creator Jason Japaljarri Woods provided the artwork (including the trees, tyre tracks and the alphabet), while exhibition designers Arketype created a dynamic graphic approach and colourful structure which would also travel easily. The almost chaotic blend of sounds, video and 'things to push and prod' complement the exhibition's real stars: two original cars from the show.

THE CARS

A number of cars appeared in *Bush Mechanics*, but the two most fondly remembered by fans of the series are the blue Holden EJ Special Station Sedan from the first episode of the series, and the spectacular painted Ford ZF Fairlane from the finale. Luckily, both were easy to locate as they had been collected by large Australian public collecting institutions shortly after the filming of *Bush Mechanics*. Requests to borrow the Holden and the Ford for a touring exhibition were received enthusiastically by the National Museum of Australia and Museums Victoria respectively. Attempts were also made to locate another of the show's vehicles, a green Holden VB Commodore that featured in the original documentary. This episode showed many *Bush Mechanics* tricks, and the VB Commodore was an accommodating and resilient test subject, but its whereabouts remain unknown. It is likely that, as with other 'old bombs' that avoided a quiet retirement in a museum collection, it continued to be repaired and enjoyed until finding a final resting place, when it became an ensemble of handy spare parts.

1962/3 HOLDEN EJ SPECIAL STATION SEDAN

The EJ is a Holden model from the booming postwar period when sales of locally built cars made up a large proportion of total car sales in Australia and when new Holden models were developed frequently and received enthusiastically by drivers. This particular EJ belonged to Francis Jupurrurla Kelly himself, who was very eager for it to feature on screen. The first

few minutes of the four-part series show the Jupurrurlas finding the parts missing from the car, which is resting in the police car yard. 'It's a piece of rubbish' declares Stephen, obviously less fond of it than Francis. Finding spares for popular models of Holdens and Fords is not particularly problematic in Yuendumu, however, and in the exhibition there is a display case by the

The EJ Holden in the *Bush Mechanics* series. National Film and Sound Archive of Australia 135784.

Francis Jupurrurla Kelly's signature. *Bush Mechanics: The exhibition*. Photographer: Andre Castellucci. National Motor Museum.

car that features an image of the Yuendumu dump and some of the badges from the hundreds of cars – mainly Fords, Holdens and the odd Toyota – that rest in this 'parts paradise'.

Some of the bush mechanics' most memorable fixes were carried out on the EJ Holden, including the replacement of a crossmember with a mulga branch. But the ingenuity of the bush mechanics is captured most arrestingly by the EJ's 'cabriolet conversion'. The car's roof caved in under the weight of gear being lugged across the bumpy road between Yuendumu and Willowra, where the bush mechanics were set to play a concert for the local children. The roof was promptly hacked off with an axe and flipped over to be towed as a trailer, its underside loaded with all the instruments and amplifiers previously strapped on top of it. Preserved just as it was left after the show's filming, the Holden illustrates a piece of lateral thinking that would hardly cross the minds of the millions of city-dwellers who enjoyed watching *Bush Mechanics* on television, and is considered by many to capture in full the spirit of *Bush Mechanics*.

Previous spread: The EJ Holden on display. *Bush Mechanics: The exhibition*. Photographer: Andre Castellucci. National Motor Museum.

1972 FORD ZF FAIRLANE 'NGAPA CAR'

A disbelieving chortle can often be heard when somebody sees the 'Ngapa Car' for the first time. Accustomed as we are to the homogeneity of modern automobiles, in which welcome advances in safety standards have rather stifled the creative impetus of design teams, the *Bush Mechanics* Ford Fairlane is a truly unique car. The star of the series' last episode, it is covered entirely in a painted *kuruwarri* (the iconographic elements) of a *Ngapa Jukurrpa* or Water Dreaming by its custodian, Thomas Jangala Rice. The episode's plot sees the bush mechanics receive the car from Jangala and drive it to Broome, to give it to an old man called Bamba in return for pearl shells. This is a recreation of ancient trading activities that brought precious pearl shells to the Central Desert for use in rain ceremonies. As director David Batty recalled, after Jangala performed his ceremony there were such violent downpours, after months of drought, that some nearby communities had to be relocated due to the devastation brought on by the storm.

The 'Ngapa Car' is accompanied by an augmented reality app that explains the *Jukurrpa* with the help of Thomas Jangala Rice and his son Donovan. *Bush Mechanics: The exhibition.* Photographer: Andre Castellucci. National Motor Museum.

Following spread: The 'Ngapa Car' in the *Bush Mechanics* series. National Film and Sound Archive of Australia BM 360.

The 'Ngapa Car' gives us insights into the continuing importance of *Jukurrpa* (Dreaming) for Warlpiri people today. A *Jukurrpa*'s importance lies in its role in the transmission of knowledge and skills through generations, and this particular *Ngapa Jukurrpa* illustrates the skill of the Warlpiri in controlling rainfall through fire. Large fires can be lit to propitiate *mangkurdu*, or pyrocumulus clouds, which in turn can become storm clouds and cause flooding when the desert is in need of rainfall. The snaking lines that run down the 'Ngapa Car' represent these floodwaters, which gather in soakages shown on its boot. The duck prints on the car's boot symbolise the animals that flock to soakages and provide food for Warlpiri hunters. These significant aspects of desert life are introduced discretely through this and all other episodes of *Bush Mechanics*. The show's ability to convey more than the motoring life of Warlpiri people is behind its success in captivating audiences of all ages and walks of life.

The Fairlane on display. *Bush Mechanics: The exhibition*. Photographer: Andre Castellucci. National Motor Museum.

THE CLAYMATION

Nestled in a case on a floor of red sand are six plasticine sculptures that resemble some of the live-action characters in *Bush Mechanics*. One holds a battery, another a gun, and another is opening a can of drink while missing a shoe. A small orange plasticine car that was used for long shots is also in the case, dwarfed by the seated elder beside it. They are the original figures used to create the *Bush Mechanics* claymation by PAW Media duo Jason Japaljarri Woods and Jonathan Daw. The claymation is faithful to the spirit of the live-action series, but Daw's previous work at Aardman Studios (creators of *Chicken Run*, *Wallace and Gromit*, and *Shaun the Sheep*) is also identifiable in the playful and upbeat style of the claymation. This eleven-minute episode, screened in its entirety in the exhibition, was produced in 2015 after a successful crowdfunding campaign. While its tone and tempo immediately attract viewers of all ages, the claymation was made specifically to be enjoyed by Warlpiri children who cannot view the original live-action *Bush Mechanics* episodes. Like many other Aboriginal groups, Warlpiri do not allow representations of the dead (including moving or still images, recordings of their voices or even use of their names). The claymation allows Warlpiri children who were too young to enjoy the series before some of its cast died to be introduced to *Bush Mechanics* tricks.

The *Bush Mechanics* claymation. Photographers: Jonothan Daw and Jason Japaljarri Woods. Pintubi Anmatjere Warlpiri Media.

Following spread: The claymation figures on display. *Bush Mechanics: The exhibition*. Photographer: Andre Castellucci. National Motor Museum.

Model mechanics

These figures sculpted by Jason Japaljarri Woods and Jonathan Daw were used in the filming of the *Bush Mechanics* claymation. You can watch the entire claymation in the exhibition's theatrette.

PAW Media Collection

Jonathan Daw working on the claymation. Pintubi Anmatjere Warlpiri Media.

Jason Japaljarri Woods working on the claymation. Pintubi Anmatjere Warlpiri Media.

Francis Jupurrurla Kelly and the other Jupurrurlas learned these tricks from their fathers, cousins, uncles. Along with the senior men and women on the board of PAW Media, they hope the claymation will instill in the young generation of today a desire to grow up as bush mechanics. If the entranced expression of many delighted young exhibition visitors watching the claymation is anything to go by, it was a winning move.

A WIDER FRAME

On first impression, *Bush Mechanics* appears to be set entirely in a 'bloke's world'. Not only are the stars of the production all young men, but most of the minor characters encountered during the series were also rather fit young men, with the exception of a few senior men. The exhibition presents visitors with a fuller picture by presenting motoring in an Aboriginal desert community outside the *Bush Mechanics* bubble. A *tin turuki*, a popular toy in Central Australia made from a discarded milk tin and fencing wire, shows how children begin their path toward becoming bush mechanics long before driving real cars, with their very own 'truckies' to break and mend. A *coolamon* (a vessel for carrying food or drink) from Warburton in Western Australia, now in the collection of the South Australian Museum, is on display to demonstrate the effect of motoring on domestic life – it is, in fact, a reshaped hubcap from an old Volkswagen. But the advent of the car in the desert has had a much more far-reaching and profound impact on life for Aboriginal Australians, as described richly in Georgine Clarsen's earlier chapter.

THE TOUR

With the support of the federal government's Visions of Australia program, which funds tours of culturally significant objects around Australia, the tour of *Bush Mechanics: The exhibition* covers thousands of kilometres. After being first displayed at the National Motor Museum in Birdwood, its next stop is the birthplace of *Bush Mechanics*, the small community of Yuendumu. Appropriate in terms of the partnership between the museum and Pintubi Anmatjere Warlpiri Media, it is also a form of short-term cultural repatriation, returning the Holden and the 'Ngapa Car' to the place that transformed them from used motor cars to nationally recognisable objects of sufficient significance to be in the collections of the National Museum of Australia and Museums Victoria. The tour's itinerary then follows the Stuart Highway to Tennant Creek, another important centre for Warlpiri and other Aboriginal groups in Central Australia, before a stop at the 2017 Royal Darwin Show. The National Road Transport Hall of Fame in Alice Springs follows, and then a return to South Australia for an Adelaide season as part of the Tarnanthi Festival of Contemporary Aboriginal and Torres Strait Islander Art. The tour concludes at the institutions that lent the original cars: Melbourne Museum, and finally, the National Museum of Australia. The interest in *Bush Mechanics*, however, has been so strong that other venues throughout Australia may follow, from northern Western Australia to southern Tasmania. The response to the latest iteration of a project begun by a small team of Aboriginal and non-Aboriginal Australians in the remote Tanami Desert in the late 1990s, armed with a small budget, some broken-down cars, an antique camera and generations of accumulated ingenuity, is remarkable.

A LIFE THAT NEVER SEEMS TO END

Director David Batty once said of *Bush Mechanics*, 'it just has a life that seems to never end'. As a testament to its enduring popularity, *Bush Mechanics* is set to return to screens. Pintubi Anmatjere Warlpiri Media are currently developing a pilot for *Bush Mechanics: Next generation*. At the same time, *Bush Mechanics: The exhibition* is taking the show to a new audience in a new context – the museum, allowing visitors to get a more intimate look at the series' cars and characters. Along the way, the characters, their skills and humour, are being seen by thousands of people, many for the first time. Perhaps they have inspired Australia's next generation of bush mechanics.

Michelangelo Bolognese is a senior curator at the National Motor Museum, and has previously worked at the British Museum and London Fire Brigade Museum. Mandy Paul is the director of the Migration Museum, another museum of the History Trust of South Australia, and has previously worked in museums around Australia and in the United Kingdom. She has a long history of working in Aboriginal Australia, which began when she was native title historian at the Central Land Council, Alice Springs.

BUSH MECHANICS